Plant Secondary Metabolites - Occurrence, Structure and Role

Edited by Alfredo Aires

Published in London, United Kingdom

Plant Secondary Metabolites - Occurrence, Structure and Role
http://dx.doi.org/10.5772/intechopen.1005930
Edited by Alfredo Aires

Contributors
Akanksha Singh, Anshul Watts, Archana Watts, Arturo Cano-Flores, Ashish Mohanty, Ashwini Kumar Dixit, Babyrani Panda, Bibi Rohida, Debashis Mahapatra, Dipsikha Mohanty, Driss Ousaaid, Gyanisha Nayak, Hassan Laaroussi, Ilham El Arabi, Jajati Keshari Nayak, Jeetendra Senapaty, Kundan Ojha, Priyanka Kumari, Satish Dubey, Shruti Sinha, Subhankar Mondal, Sushree Sangeeta, Yuvonne Angel Lyngdoh

First published in London, United Kingdom, 2025 by IntechOpen
IntechOpen is the global imprint of INTECHOPEN LIMITED, registered in England and Wales, registration number: 11086078, 167-169 Great Portland Street, London, W1W 5PF, United Kingdom

For EU product safety concerns: IN TECH d.o.o., Prolaz Marije Krucifikse Kozulić 3, 51000 Rijeka, Croatia, info@intechopen.com or visit our website at intechopen.com.

British Library Cataloguing-in-Publication Data
A catalogue record for this book is available from the British Library

Plant Secondary Metabolites - Occurrence, Structure and Role
Edited by Alfredo Aires
p. cm.
Print ISBN 978-1-83634-545-9
Online ISBN 978-1-83634-544-2
eBook (PDF) ISBN 978-1-83634-546-6

Meet the editor

Alfredo Aires is an Associate Professor in the Department of Agronomy at the School of Agrarian and Veterinary Sciences, UTAD, Portugal. He holds a degree in Agronomic Engineering, a Master's in Crop Science with a specialization in Horticulture and Ornamental Plants, and a PhD in Agronomic Sciences. Since 1996, he has combined teaching and research, participating in 12 national and international R&D projects, including one as national PI under the ERA-Net program. A full member of CITAB/UTAD and INOV4Agro, he coordinates research on local food systems. He serves on the editorial board of *Industrial Crops and Products* (Elsevier) and is a member of SCAP, APH, and ISHS. His research focuses on horticultural and arable crop production, nutrient composition, and bioactive compounds.

Contents

Preface

Plants produce a vast array of secondary metabolites, compounds not essential for basic growth, yet vital for defense, adaptation, and interaction with their environment. These substances, including alkaloids, terpenoids, and phenolics, are equally important to humans, providing medicines, nutraceuticals, flavors, and agricultural innovations. This book explores their occurrence, structural diversity, biosynthesis, and roles in plant ecology, while linking fundamental knowledge to practical applications in health, agriculture, and industry. Designed as a clear and comprehensive reference, it serves students, researchers, and professionals interested in the biological importance and real-world potential of these fascinating natural products.

Alfredo Aires
CITAB - Centro de Investigação e Tecnologias Agroambientais e Biológicas,
Inov4Agro - Instituto para a Inovação, Capacitação e Sustentabilidade
da Produção Agroalimentar,
Universidade de Trás-os-Montes e Alto Douro (UTAD),
Vila Real, Portugal

Departamento de Agronomia,
Escola das Ciências Agrárias e Veterinárias,
Universidade de Trás-os-Montes e Alto Douro (UTAD),
Vila Real, Portugal

Section 1

Classification, Structure and Biosynthesis

Chapter 1

The Classification of Plant Bioactive Compounds, their Structure and Applications in Daily Life

Bibi Rohida

Abstract

Plant secondary metabolites are diverse chemical compounds that are not directly involved in the basic metabolic processes of plants but play crucial roles in plant defense, signaling, and ecological interactions. These metabolites are typically categorized into three main groups: alkaloids, terpenoids, and phenolics, each with unique structures and functions. Alkaloids include nitrogen-containing compounds such as caffeine and morphine, which are often toxic to herbivores and have medicinal properties. Terpenoids, the largest group, encompass essential oils, carotenoids, and steroids, contributing to plant defense mechanisms, aroma production, and pigmentation. Phenolic compounds, including flavonoids, tannins, and lignans, are known for their antioxidant properties and involvement in plant protection against UV radiation, pathogens, and herbivores. These secondary metabolites have significant economic importance due to their roles in pharmaceuticals, agriculture, and the food industry. The study of plant secondary metabolites and their biosynthetic pathways is crucial for developing sustainable approaches to harness these compounds for various industrial applications, including drug discovery, pest control, and food preservation. This review aims to provide a comprehensive overview of the major classes of plant secondary metabolites, their biological significance, and their importance in biotechnology and human health.

Keywords: phytochemicals, terpenoids, phenolic compounds, alkaloids, saponins antioxidants

1. Introduction

1.1 Metabolites

The intermediate products of metabolism, known as metabolites, are produced by a variety of naturally occurring enzymes in cells, such as pigments and antibiotics. Small molecules are typically referred to as metabolites. Metabolites serve a variety of purposes, such as defense, structure, signaling, fuel, catalysis, and interactions with other living things. Humans, microbes, and plants all produce metabolites. Metabolites are the final or intermediate products formed during metabolic processes. They are divided into two types: primary metabolites and secondary metabolites [1, 2].

1.1.1 Primary metabolites

These are the chemical substances that are produced throughout the processes of growth and development. Additionally, they participate in photosynthesis and respiration, the two main metabolic processes. The growth phase is when the primary metabolites are produced. They are referred to as central metabolites and are responsible for maintaining the plant physiological processes. These are the byproducts of anabolic metabolism that are utilized by cells to create necessary macromolecules. Among the main metabolites produced industrially are organic acids, vitamins, and amino acids. The main primary metabolite that is produced industrially and on a large scale is alcohol.

1.1.2 Secondary metabolites

Secondary metabolites are organic compounds that are not directly involved in the normal growth, development, or reproduction of plants but play crucial roles in their survival and adaptation. Secondary metabolites are used by humans as recreational drugs, flavorings, pigments, and medications [3]. Plants are autotrophic organisms. All living things have a secondary metabolism in addition to their primary metabolism, which enables them to create and accumulate a wide variety of chemical compounds [4].

Secondary metabolites (SMs) are organic substances that plants create and are commercially employed across a range of agriculture, cosmetics, and pharmaceutics, among other sectors [5, 6]. Human health and well-being are tremendously aided by these metabolites, as they exhibit a strict range of biological functions. Such functions include antibacterial, antioxidant, and anti-cancer activities [7, 8]. The relative frequency with which a variety of SMs is used in the treatment of diseases is assumed to be the reason for the use of SMs in the form of medicinal plants. These useful substances are, as a rule, present in all living organisms, and as a mechanism of adaptation the plant has to biosynthesize them under stressful conditions. Genetic factors, climate and food supply are among the various determinants which can influence the capacity of plants to synthesize these compounds.

Plants are subjected to several environmental challenges which include pathogenic (viruses and nematodes, fungi, and insects), physical (salinity, temperature, drought, metal, and UV), and so on. The stressors are "seen" by the receptors on the plant plasma membrane, which set off a chain of events leading to the biosynthesis of SMs [5, 6].

Plant SMs can be broadly classified into four groups: alkaloid terpenes, flavonoids, phenolic, and steroids. These substances serve a variety of purposes, such as altering symbiotic signaling, herbivore repulsion, and the organization of microbial communities [9–12].

1.2 Historical overview

1. *Ancient uses*: The use of plants for medicinal purposes dates back to ancient civilizations. Early herbalists recognized the importance of certain plant compounds, often relying on trial and error to identify those with beneficial effects.

2. *Eighteenth and nineteenth centuries*: The scientific study of secondary metabolites began to flourish during this period. Botanists and chemists started isolating and

characterizing compounds like alkaloids (e.g., morphine from opium poppy) and glycosides.

3. *Twentieth-century advances*: With the development of modern chemistry and analytical techniques, researchers were able to isolate, purify, and study the structure and function of various secondary metabolites. This era saw significant discoveries, such as the identification of flavonoids [13], terpenoids, and tannins.

4. *Ethnopharmacology*: The late twentieth century emphasized the importance of traditional knowledge and practices in identifying plant secondary metabolites with medicinal properties. This led to a resurgence in the study of natural products and their pharmacological effects.

5. *Biotechnology and synthetic biology*: In recent decades, advances in biotechnology have allowed for the genetic manipulation of plants to enhance the production of valuable secondary metabolites. Techniques such as metabolic engineering and synthetic biology have opened new avenues for producing compounds like artemisinin (used in malaria treatment) (**Table 1**).

Some ancient communities have benefited from the use of these compounds in traditional medicine, despite their ignorance of them, for instance, applying turmeric paste to wounds or making a paste out of the stem of the neem plant. Traditional Chinese medicine has made use of the herb *Artemisia annua*. Secondary compounds continue to play a role in modern people's lives. It has been widely utilized in food additives, medicines, agrochemicals, and aromatics.

The secondary metabolites are divided into five main classes according to the compound's structures.

- terpenoids,
- phenolics,
- polyketides,
- alkaloids,

Primary metabolites	Secondary metabolites
Required for growth and maintenance of cellular function	Involved in ecological functions
Occurs at the growth phase	Occurs at the stationary phase
Produced in large amounts and easy to extract	Produced in small amounts and difficult to extract
Same in every species	Different in every species
Perform physiological functions in the body	Derivatives of primary metabolites
For example, carbohydrates, vitamins, ethanol, lactic acid	For example, phenolics, steroids, antibiotics, pigments

Table 1.
Difference between primary and secondary metabolites.

- carbohydrates,
- glycosides,
- saponins, and
- essential oils.

2. Terpenoids

This group of secondary metabolites is broad and varied. Additionally, it has a few key metabolites that are necessary for the growth and development of plants. Terpenes are simply lengthy chains of hydrocarbons, whereas terpenoids are oxygenated hydrocarbons. The general formula for it is $(C_5H_8)_n$, where n is the number of units of 5-carbon isoprene. An alternative name for them is isoprenoid compounds. The number of isoprene units in the terpenoids determines their classification. It consists of:

Hemiterpene: It consists of a single isoprene unit. Its examples include isovaleric acid from *Vaccinium myrtillus* and angelic acid isolated from *Angelica archangelica*.

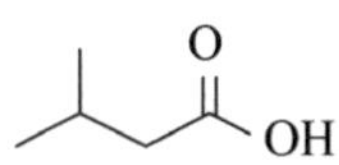

isovaleric acid

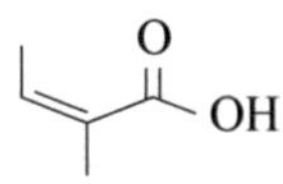

angelic acid

Monoterpene: It consists of two isoprene units. They are further classified based on their structure as hydrocarbons, alcohol, ketones, alcohol esters, and aldehydes. Its examples are Carvone, Linalool, Limonene, Linalyl acetate, and Citronellal.

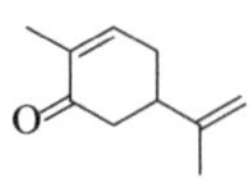

Carvone

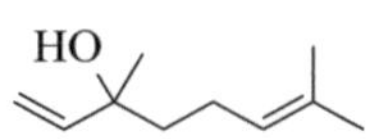

Linalool

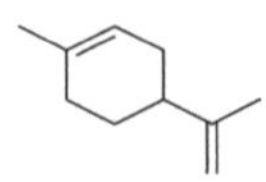

Limonene

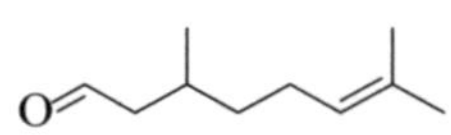

Citronellal

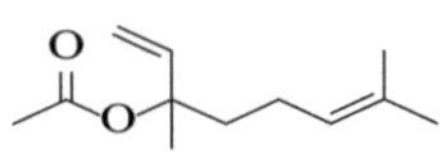

Linalyl acetate,

Sesquiterpene: It has three isoprene units. Its examples include caryophyllene and farnesol. A number of these compounds show antibacterial, antiprotozoal, and antifungal activities.

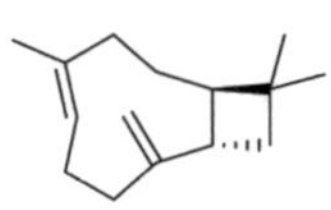

caryophyllene

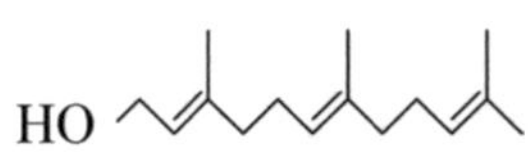

farnesol.

Diterpene: It contains four isoprene units. A few examples of this group are gibberellins, cembrane, phytane, and labdane. They possess a range of medicinal activities, including antifungal, antibacterial, analgesic, anti-inflammatory, and antineoplastic activities.

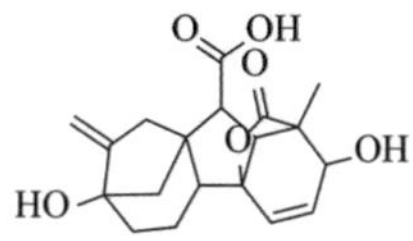

gibberellins

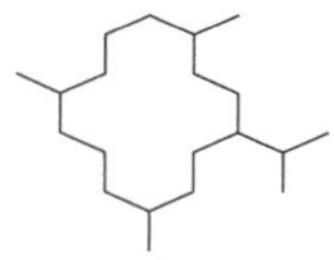

cembrane

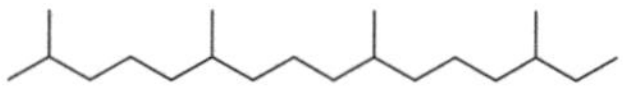

phytane

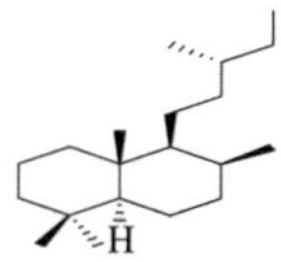

labdane

Sesterterpenes: It consists of five isoprene units. An example is geranyl farnesol.

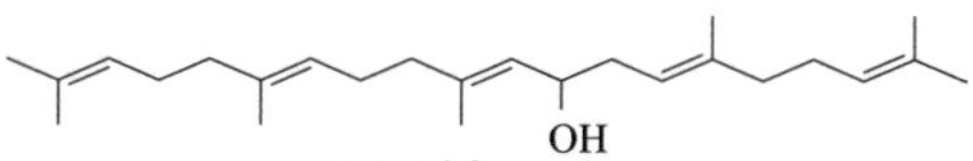

geranyl farnesol

Triterpenes: It has six isoprene units. An example of triterpene is quassin.

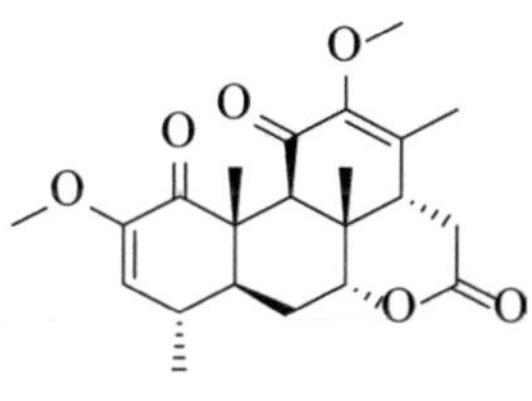

Quassin

Sesquarterterpene: It is composed of seven isoprene units.

Tetraterpene: It contains eight isoprene units. Its examples are carotenoids and xanthophylls.

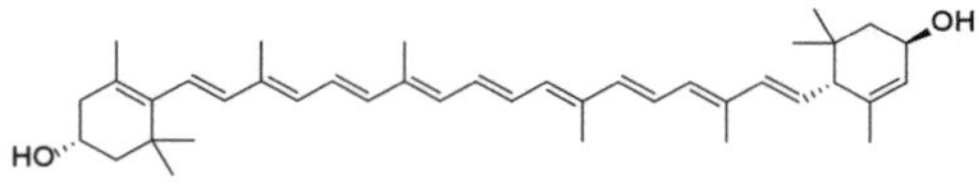

Carotenoids

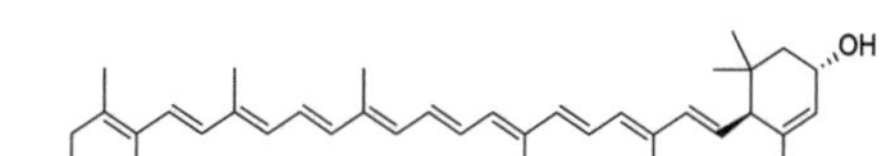

Xanthophylls

Polyterpene: It includes molecules having more than eight isoprene units. An example of this group of terpene is natural rubber.

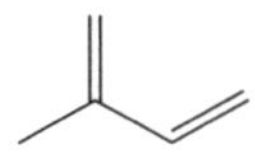

Natural rubber

2.1 Applications across various fields

1. *Pharmaceuticals*: Many terpenoids possess therapeutic properties. For example, paclitaxel (from the Pacific yew tree) is used in cancer treatment, while artemisinin (from sweet wormwood) is used to treat malaria.

2. *Fragrances and flavorings*: Terpenoids contribute to the scent and flavor of many essential oils. Compounds like limonene (found in citrus) and linalool (in lavender) are widely used in perfumes and food products.

3. *Agriculture*: Some terpenoids have insecticidal and antifungal properties, making them useful as natural pesticides. They can help protect crops from pests and diseases.

4. *Cosmetics*: Terpenoids are often included in cosmetic formulations for their aromatic qualities and potential skin benefits. Ingredients like tea tree oil and eucalyptus oil are popular in skincare.

5. *Food industry*: Beyond flavoring, terpenoids are used as natural preservatives due to their antioxidant properties.

6. *Biofuels*: Certain terpenoids can be used as biofuels or additives in biofuels, offering a renewable energy source.

7. *Industrial applications*: Terpenoids are utilized in the production of resins, adhesives, and plastics. For example, terpene resins are used in printing inks and coatings.

8. *Research*: Terpenoids are subjects of extensive research for their potential in developing new drugs and understanding ecological interactions among plants and animals.

3. Phenolics

These are organic compounds with one or more hydroxyl (-OH) groups and an aromatic ring structure. They are the biggest and most prevalent class of secondary metabolites in plants. From a single aromatic ring to a highly complex polymeric compound, they have a variety of structures. Phenolic compounds are often considered as the predominant class of secondary metabolites. Phenols, or polyphenols, are characterized by their small size and aromatic nature. These compounds are synthesized by either the polyketide acetate/malonate route or the shikimate/phenylpropanoid

DOI: http://dx.doi.org/10.5772/intechopen.1010332

system. They play crucial roles in several physiological processes, as highlighted by Naikoo et al. [14].

They are categorized according to their structure and place of biosynthesis. Based on their structural makeup, they are categorized as follows:

Simple phenolics: They are defined as compounds having at least one hydroxyl group attached to the basic skeleton of an aromatic ring. It includes compounds like gallic acid, eugenol, catechol, salicylic acid, hydroquinone, and thymol.

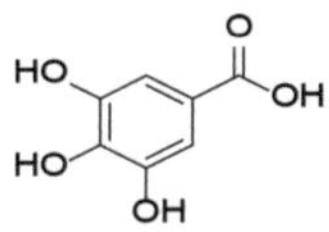
gallic acid,

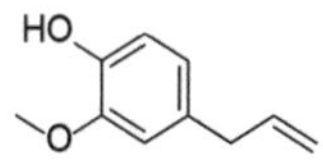
eugenol

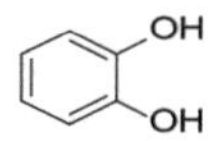
catechol

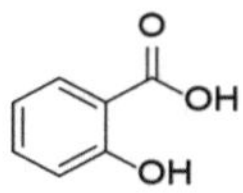
salicylic acid

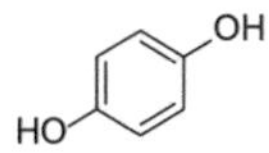
hydroquinone

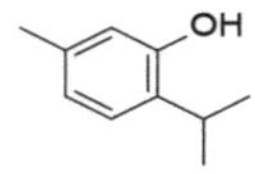
thymol.

Coumarins: It is a derivative of benzo-α-pyrone. Clover and melotot are the richest sources of coumarins. Its examples are esculetin, scopoletin, and umbelliferone.

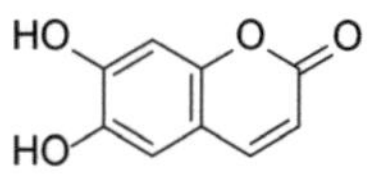
esculetin

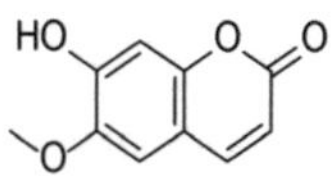
scopoletin

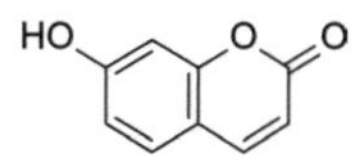
umbelliferone.

Tannin: These are water-soluble phenol derivatives. They are of two types: hydrolyzable tannins and condensed tannins. Its examples include gallotannins, geraniin, and ellagitannins.

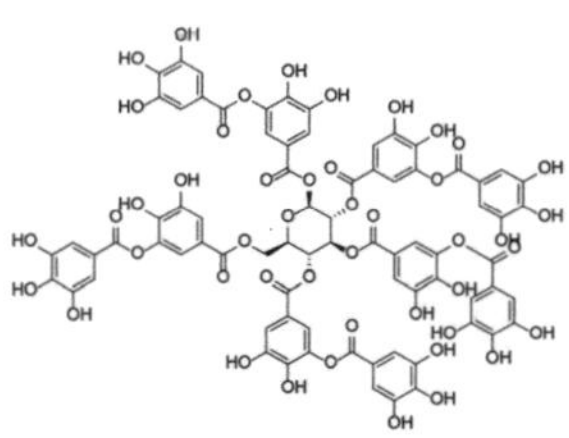
Gallotannins

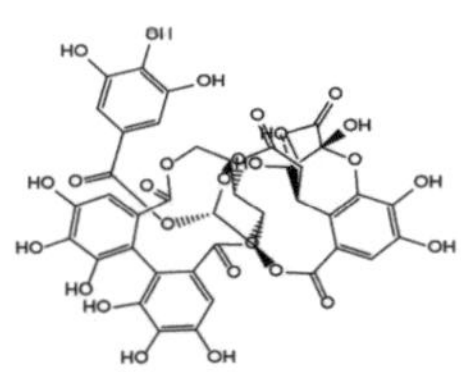
geraniin

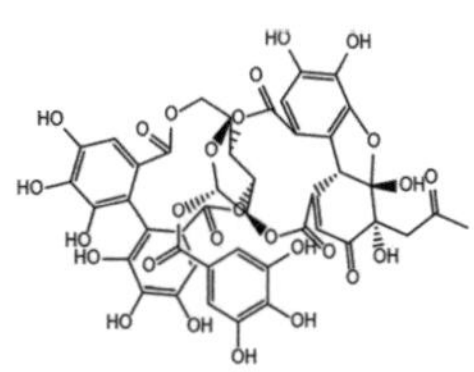
ellagitannins.

Flavonoids: These are the largest group of phenolics secondary metabolites, found in almost all vegetables and fruits. In plants, flavonoid aglycones occur in a variety of structural forms. Polyphenols, or flavonoids (C_6-C_3-C_6), generally consist of two benzene rings linked together in a chain, namely ring A and ring B, which consist of three carbons. Usually, this string is condensed into one pyranic cycle, called the C cycle. The C-ring of this connection consists of flavones, flavonols, flavanones, flavanonols, flavanols, catechins, and anthocyanins [15].

It is predominantly found in the Umbelliferae, Polygonaceae, Leguminosae, Compositae, and Rutaceae families of plants.

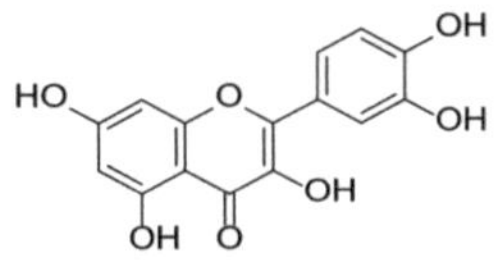

quercetin

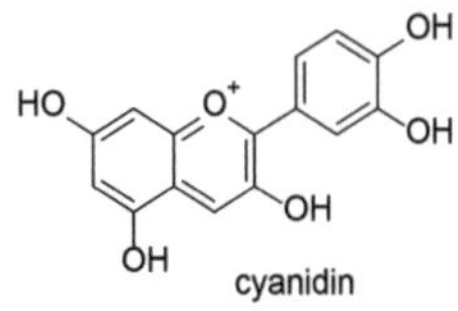

cyanidin

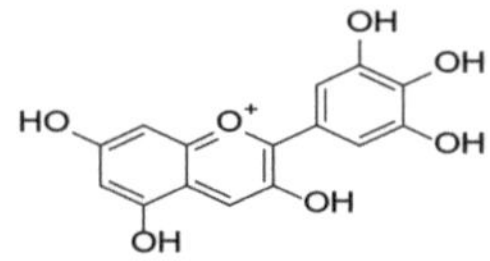

delphinidin

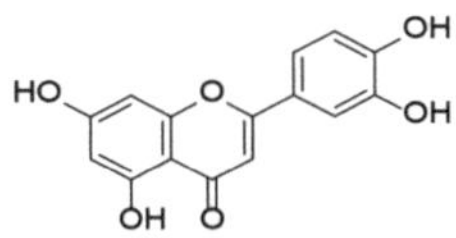

luteolin

Lignans: They are dimeric compounds formed by joining two molecules of phenyl-propene derivative. Their examples include matairesinol and Wikstrom.

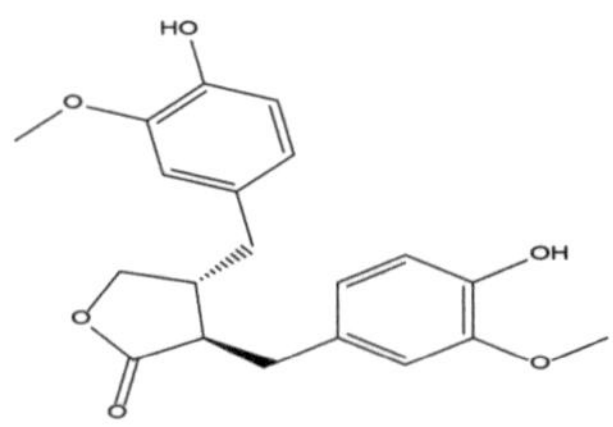

Matairesinol

Stilbenes: They are a widely distributed small group of phenolic compounds. Its example is resveratrol.

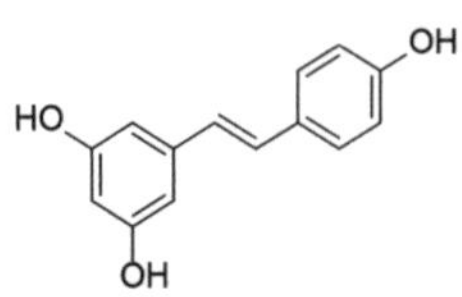

resveratrol.

Chromones and xanthones: These are structural derivatives of benzo-γ-pyrone. They are not of much pharmaceutical importance. Their examples are eugenin, khellin, and furanochromones.

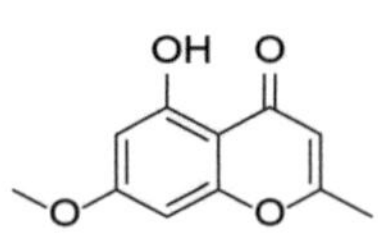

Eugenin

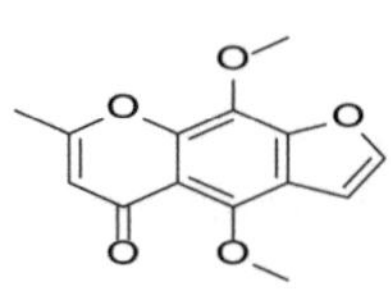

Khellin

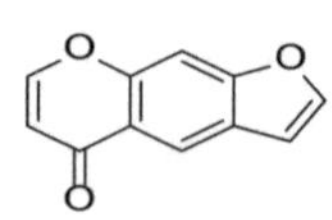

furanochromones.

3.1 Applications

1. *Pharmaceuticals*: Many phenolic compounds have antioxidant, antimicrobial, and anti-inflammatory properties, making them valuable in drug development.

2. *Food industry*: Phenolic compounds are used as preservatives and flavoring agents. They are also important for their antioxidant properties, which help in food preservation.

3. *Plastics and resins*: Phenolic resins are widely used in the production of molded products, laminates, and adhesives. They are known for their thermal stability and chemical resistance.

4. *Cosmetics*: Many phenolic compounds are used in skincare products for their antioxidant properties and ability to improve skin health.

5. *Agriculture*: Certain phenolics act as natural pesticides and herbicides, contributing to plant defense mechanisms against pests and diseases.

6. *Construction materials*: Phenolic compounds are used in wood preservation and in the production of fiberboards and insulation materials.

7. *Biomaterials*: In the biomedical field, phenolics are explored for their potential in tissue engineering and drug delivery systems.

8. *Dyes and pigments*: Some phenolic compounds are used to produce dyes and pigments in textiles and coatings.

4. Poly ketides

Polyketides are a diverse class of natural products characterized by their complex structures and significant biological activities. They are produced by a variety of organisms, including bacteria, fungi, and plants, and they play important roles in ecological interactions and human medicine.

4.1 Structure and biosynthesis

Polyketides are synthesized through the condensation of acetyl-CoA and other acyl-CoA units via a process similar to fatty acid biosynthesis. The key features of polyketide biosynthesis include:

1. *Polyketide synthesis (PKS)*: These are large multi-enzyme complexes responsible for the assembly of polyketide chains. There are two main types of PKS:

    - *Type I PKS*: These are modular and operate in a "one-pot" manner, where all the enzymes are part of a single polypeptide chain.

- *Type II PKS*: These are iterative and consist of separate enzyme components that work together sequentially.

2. *Decarboxylation and modification*: After the initial chain is formed, various modifications such as methylation, hydroxylation, and glycosylation can occur, leading to the final structure of the polyketide.

4.2 Types of polyketides

Polyketides can be classified based on their structure and biological activity:

1. *Aromatic polyketides*: These include compounds with aromatic rings, such as:

- *Antibiotics*: For example, erythromycin and tetracycline.

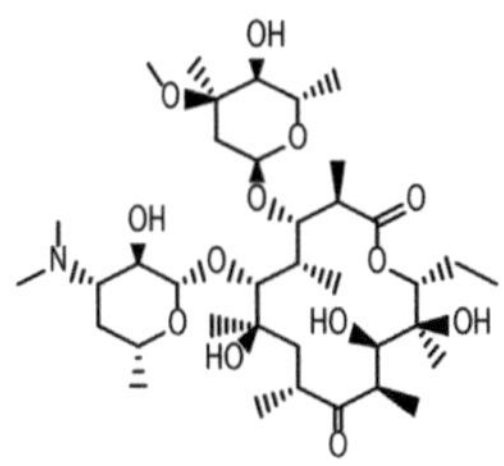

Erythromycin

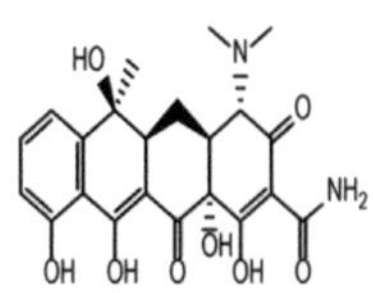

tetracycline

- *Pigments*: For example, anthraquinones.

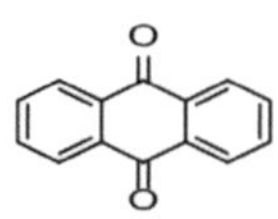

anthraquinones

2. *Aliphatic polyketides*: These consist of long carbon chains and can be further divided into:

- *Fatty acids*: For example, mycolic acids are found in the cell walls of mycobacteria.

- *Lactones*: For example, polyene macrolides like nystatin.

3. *Macrocyclic polyketides*: These are large, ring-structured compounds with significant pharmacological properties, such as:

- *Immunosuppressants*: For example, rapamycin and FK506.

4. *Polyene antifungals*: Such as amphotericin B, which target fungal membranes.

5. *Polycyclic aromatic compounds*: Compounds like avermectin, which are used in agriculture and medicine.

4.3 Applications

Polyketides have important applications in various fields:

Pharmaceuticals: Many polyketides serve as antibiotics, antifungals, anti-cancer agents, and immunosuppressants.

Agriculture: Some polyketides are used as pesticides and herbicides due to their biological activity against pests and diseases.

Biotechnology: Polyketides are studied for their potential in synthetic biology, aiming to engineer microorganisms to produce novel compounds.

Alkaloids are a diverse group of naturally occurring organic compounds that mostly contain basic nitrogen atoms. They are primarily found in plants and can have profound effects on humans and animals, often influencing physiological processes. Here is a comprehensive overview of alkaloids, their characteristics, and types:

4.4 Characteristics of alkaloids

1. *Basic nature*: Most alkaloids are basic due to the presence of nitrogen, which can accept protons.

2. *Bitter taste*: Many alkaloids have a bitter flavor, which can deter herbivores from eating the plants.

3. *Physiological effects*: Alkaloids can have significant effects on the nervous system, cardiovascular system, and more, often used in medicine.

4. *Complex structures*: They can have complex ring structures and can be derived from various amino acids.

5. *Solubility*: Alkaloids are generally soluble in organic solvents and less soluble in water.

4.5 Types of alkaloids

Alkaloids can be categorized based on their chemical structure and the plant sources from which they are derived. Here are the main categories:

4.5.1 Classification by structure

Pyridine and piperidine alkaloids: These contain a six-membered ring with nitrogen. Examples include nicotine (from tobacco) and piperine (from black pepper).

Quinoline and isoquinoline alkaloids: These contain a fused ring structure. Examples include quinine (from cinchona bark) and morphine (from opium poppy).

Indole alkaloids: These have an indole structure. Examples include tryptamine and serotonin.

Terpenoid alkaloids: Derived from terpenes, examples include vincristine and vinblastine (from periwinkle).

Steroidal alkaloids: These have steroid-like structures. Examples include solanine (from potatoes) and toxicophylline.

4.5.2 Classification by source

Plant alkaloids: The most common source, examples include:

- *Caffeine*: Found in coffee and tea.
- *Nicotine*: Found in tobacco.
- *Morphine*: Found in opium poppy.

Animal alkaloids: Less common but can include compounds like conotoxins from cone snails.

Microbial alkaloids: Produced by fungi and bacteria, such as ergotamine from the fungus *Claviceps purpurea*.

4.5.3 Classification by effects

- *Stimulants*: Increase alertness and energy (e.g., caffeine and nicotine).
- *Sedatives*: Induce calmness or sleep (e.g., morphine and codeine).
- *Hallucinogens*: Alter perception (e.g., psilocybin from mushrooms).
- *Analgesics*: Provide pain relief (e.g., morphine and oxycodone).

The acridones, aromatics, carbolines, ephedras, ergots, imidazoles, bisindoles, indolizines, manzamines, oxindoles, quinazolines, phenyl isoquinoline, phenylethylamines, purines, pyrrolidines, pyrroloindoline, and simple tetrahydroisoquinolines are the examples of complex alkaloids.

4.6 Uses of alkaloids

Medicinal: Many alkaloids have therapeutic uses, such as morphine for pain relief and quinine for malaria treatment.

Psychoactive: Some alkaloids are used recreationally for their mind-altering effects (e.g., cocaine and LSD).

Agricultural: Certain alkaloids can be used as natural pesticides or herbicides.

4.7 Toxicity

While many alkaloids have medicinal benefits, some can be highly toxic. For example, strychnine is a potent neurotoxin, and certain alkaloids can lead to serious health issues if consumed in large quantities.

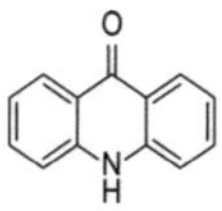
Acridones

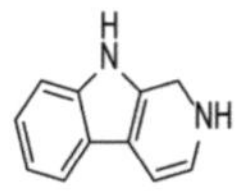
Carbolines

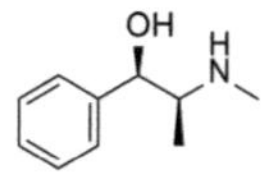
ephedrine

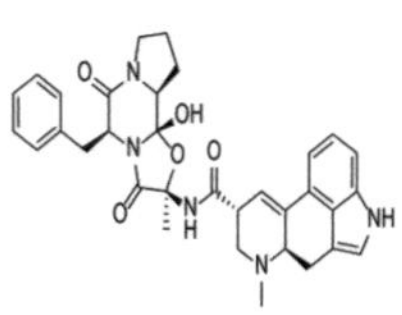
Ergotamine

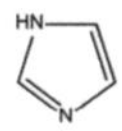
Imidazoles

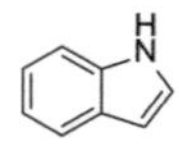
Indoles

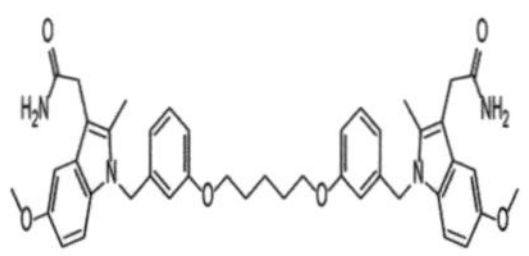
Bis-indole derivative 10

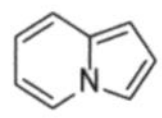
Indolizines

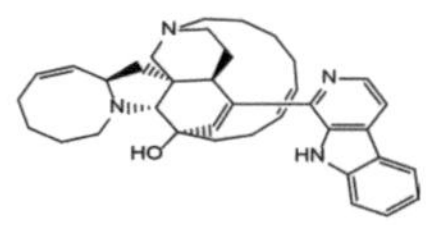
Manzamines

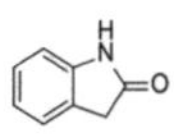
Oxindoles

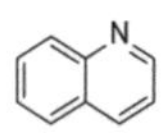
Quinolines

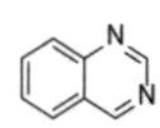
Quinazolines

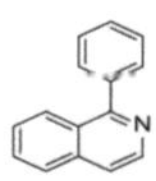
phenyl isoquinoline

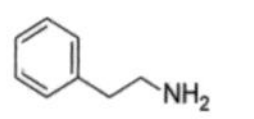
phenylethylamines

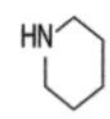
piperidines

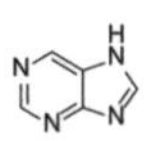
Purines

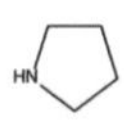
pyrrolidines

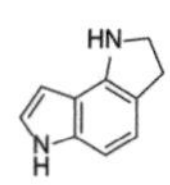
Pyrroloindoline

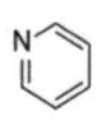
Pyridines

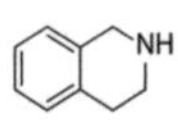
tetrahydroisoquinolines

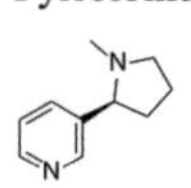
Nicotine

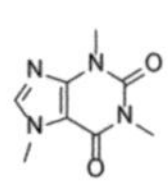
Caffeine

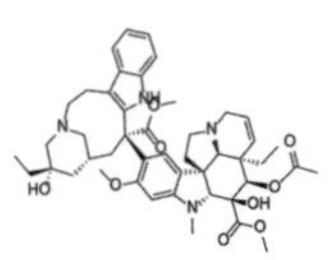
vinblastine

5. Carbohydrates

Carbohydrates are organic compounds composed primarily of carbon, hydrogen, and oxygen, and they serve as one of the main energy sources for living organisms. They can be classified into several categories based on their structure and complexity.

5.1 Classification of carbohydrates

5.1.1 Monosaccharides

These are the simplest form of carbohydrates, consisting of single sugar molecules. They cannot be hydrolyzed into simpler sugars. Common examples include:

- *Glucose*: A primary energy source for cells.
- *Fructose*: Found in fruits and honey.
- *Galactose*: Part of lactose found in milk.

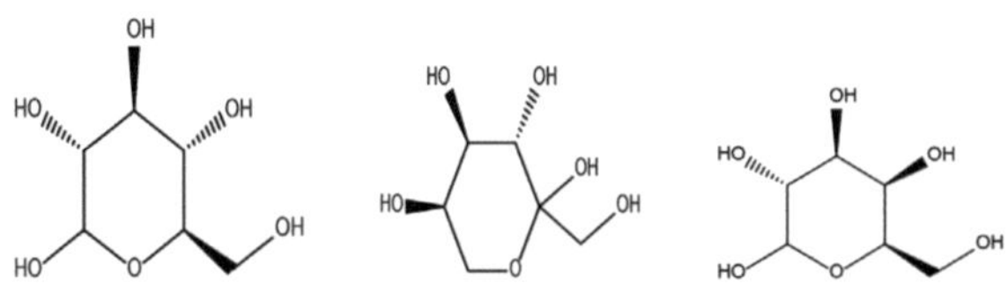

Glucose fructose galactose

5.1.2 Disaccharides

These are formed by the combination of two monosaccharides through a glycosidic bond. Examples include:

- *Sucrose*: Composed of glucose and fructose (table sugar).
- *Lactose*: Composed of glucose and galactose (milk sugar).
- *Maltose*: Composed of two glucose molecules (found in malted foods and beverages).

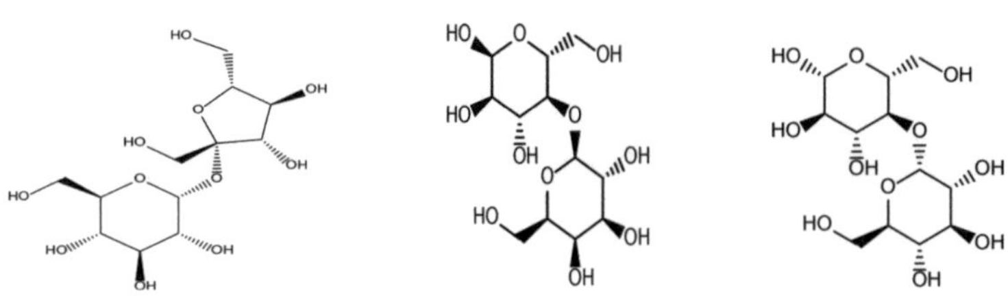

Sucrose lactose maltose

5.1.3 Oligosaccharides

These consist of 3 to 10 monosaccharide units. They are often found in beans, lentils, and certain vegetables. Common examples include:

- *Raffinose*: Composed of galactose, glucose, and fructose.
- *Melezitose*: Found in honeydew and some plant saps.

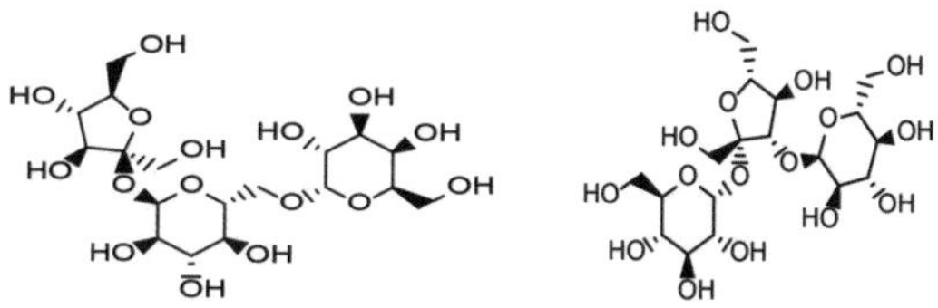

Raffinose Melezitose

5.1.4 Polysaccharides

These are complex carbohydrates made up of long chains of monosaccharide units. They can be branched or unbranched. Examples include:

- *Starch*: A storage form of glucose in plants, found in foods like potatoes and grains.
- *Glycogen*: The storage form of glucose in animals, primarily found in the liver and muscles.
- *Cellulose*: A structural component of plant cell walls, it is not digestible by humans but is important for dietary fiber.

5.2 Functions of carbohydrates

- *Energy source*: Carbohydrates provide a quick and accessible source of energy. Glucose is particularly crucial for cellular respiration.
- *Energy storage*: Starch in plants and glycogen in animals serve as energy reserves.
- *Structural role*: Cellulose provides structural integrity to plant cells, while chitin serves a similar role in the exoskeletons of arthropods and fungal cell walls.
- *Cell signaling*: Glycoproteins and glycolipids on cell surfaces play key roles in cell recognition and signaling.

5.3 Dietary sources of carbohydrates

Carbohydrates are found in a wide variety of foods:

- *Simple carbohydrates*: Fruits, honey, milk, and table sugar.
- *Complex carbohydrates*: Whole grains, legumes, starchy vegetables, and some fruits.

5.4 Digestion and metabolism

Carbohydrate digestion begins in the mouth with salivary amylase and continues in the small intestine where enzymes break them down into monosaccharides. These are then absorbed into the bloodstream and transported to cells for energy production.

5.5 Health implications

- *Balanced intake*: While carbohydrates are essential, it is important to choose complex carbs over simple sugars to maintain stable blood sugar levels and provide sustained energy.
- *Dietary fiber*: Found in fruits, vegetables, and whole grains, fiber is crucial for digestive health and can help prevent chronic diseases.
- *Excess carbohydrate consumption*: High intake of refined carbohydrates and sugars can lead to obesity, diabetes, and other metabolic disorders.

6. Glycosides

Glycosides are compounds in which a sugar (glycone) is bound to a non-sugar moiety (aglycone) through a glycosidic bond. They are widely found in plants and serve various biological functions. Here are some key points about glycosides:

1. *Structure*: The glycone part can be a monosaccharide, disaccharide, or oligosaccharide, while the aglycone can be a wide range of compounds, including phenolic compounds, terpenes, or alkaloids.

6.1 Types of glycosides

1. *Cardiac glycosides*:
    - Found in plants like foxglove (Digitalis).
    - Used in treating heart conditions by increasing the force of heart contractions.
2. *Saponins*:
    - Can form foam when shaken in water.
    - Exhibit surfactant properties and can enhance the absorption of other compounds.
    - Some have medicinal properties, while others can be toxic.
3. *Flavonoid glycosides*:
    - Important in plant coloration and protection from UV radiation.
    - Exhibit antioxidant, anti-inflammatory, and antimicrobial properties.

4. *Anthraquinone glycosides*:

- Found in plants like senna.
- Often used as laxatives due to their ability to stimulate bowel movements.

6.2 Biological functions

- *Defense mechanisms*: Glycosides can serve as a defense against herbivores and pathogens, deterring feeding or inhibiting microbial growth.
- *Transport and storage*: Some glycosides help in the storage and transport of energy within plants.
- *Signaling*: They can play roles in plant signaling pathways, affecting growth and development.

6.3 Hydrolysis

- Glycosides are often hydrolyzed in the presence of water, acids, or enzymes (like glycosidases). This reaction breaks the glycosidic bond, releasing the sugar and the aglycone. The aglycone may then exhibit biological activities distinct from the glycoside form.

6.4 Applications

1. *Pharmaceuticals*: Many glycosides are used in modern medicine. For instance:

- *Digoxin* (a cardiac glycoside) is used to treat heart failure.
- *Senna glycosides* are used in over-the-counter laxatives.

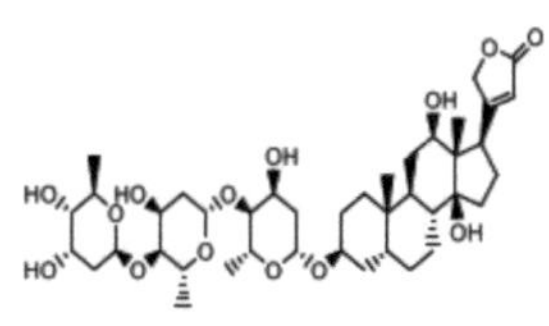

Digoxin

2. *Food industry*: Some glycosides, like steviol glycosides from stevia, are used as natural sweeteners.

3. *Cosmetics*: Glycosides can be included in skincare products for their antioxidant properties.

4. *Research*: Glycosides are studied for their potential health benefits, including anti-cancer and anti-diabetic effects.

7. Saponins

Saponins are a class of naturally occurring compounds found in various plants, known for their soap-like properties when mixed with water. They have a glycosidic structure, typically consisting of a hydrophobic aglycone (also known as a sapogenin) linked to one or more hydrophilic sugar moieties. Here is a detailed explanation of their characteristics, sources, functions, and applications:

7.1 Structure

1. *Aglycone (sapogenin)*: This is the non-sugar part of the molecule, which can be steroidal or triterpenoidal in nature.

2. *Glycosidic linkage*: The aglycone is attached to one or more sugar units, usually glucose, galactose, or rhamnose.

3. *Hydrophilic and hydrophobic regions*: The dual nature allows saponins to interact with both water and lipids, giving them surfactant properties.

7.2 Sources

Saponins are found in a wide range of plant species, including:

- *Legumes*: Such as soybeans, chickpeas, and alfalfa.

- *Roots and herbs*: Like ginseng and garlic.

- *Quinoa*: Known for its high saponin content, which can give it a bitter taste.

- *Some vegetables*: Such as spinach and asparagus.

7.3 Functions in plants

- *Defense mechanism*: Saponins can deter herbivores and pathogens due to their toxic effects.

- *Antioxidant properties*: They may protect plants from oxidative stress.

- *Root development*: Some studies suggest they play a role in root formation and growth.

7.4 Health benefits

Saponins have been associated with several health benefits in humans and animals:

- *Cholesterol reduction*: They may help lower blood cholesterol levels by binding to bile acids.

- *Immune system support*: Some saponins can enhance immune response.

- *Antioxidant effects*: They have been shown to exhibit antioxidant properties.
- *Anti-cancer potential*: Certain saponins have demonstrated the ability to inhibit cancer cell growth in laboratory studies.

7.5 Applications

1. *Food industry*: Saponins are used as natural emulsifiers and foaming agents in food products.
2. *Pharmaceuticals*: They are explored for their potential health benefits and used in some formulations.
3. *Cosmetics*: Their surfactant properties make them suitable for shampoos and other personal care products.
4. *Agriculture*: Saponins are studied for their potential as natural pesticides.

Saponins are a diverse group of glycosides found in various plants. They are characterized by their ability to produce foam when shaken in water and have both hydrophilic and lipophilic properties. Here are the main types of saponins:

1. *Triterpenoid saponins*:
    - These are derived from triterpenes and have a steroid-like structure. They are commonly found in plants like ginseng and various legumes.
2. *Steroidal saponins*:
    - These saponins are derived from steroidal compounds and are often found in plants such as Dioscorea (yams) and some species of Solanum.
3. *Oligosaccharide saponins*:
    - These contain a smaller number of sugar units attached to the aglycone part and are less common.
4. *Acylated saponins*:
    - These have fatty acids attached to the sugar moieties, enhancing their biological activity and solubility.
5. *Quinone saponins*:
    - These contain quinone structures and are less common than the other types but have interesting biological properties.

Each type of saponin may have different biological activities, including antimicrobial, antifungal, and anti-inflammatory effects, making them of interest in both traditional medicine and modern pharmacology.

7.6 Toxicity

While saponins have beneficial properties, they can also be toxic at high concentrations. Symptoms of saponin toxicity can include:

- Gastrointestinal upset (nausea, vomiting)
- Hemolysis (destruction of red blood cells)
- Liver damage in extreme cases

8. Essential oils

Essential oils are concentrated extracts derived from plants, capturing their natural fragrance and beneficial properties. They are used in aromatherapy, personal care, cleaning products, and more. Here is a detailed overview of essential oils, their types, and their uses:

- *Extraction*: Essential oils are typically extracted through methods like steam distillation, cold pressing, or solvent extraction.
- *Composition*: They contain volatile compounds that give plants their characteristic scents and can have therapeutic effects.

8.1 Types of essential oils

1. *Citrus oils*
   - *Examples*: Lemon, Orange, Grapefruit, Bergamot.
   - *Uses*: Uplifting mood, improving energy, and promoting clarity.
2. *Herb oils*
   - *Examples*: Basil, Oregano, Rosemary, Thyme.
   - *Uses*: Culinary applications, digestive support, and respiratory health.
3. *Flower oils*
   - *Examples*: Lavender, Rose, Chamomile, Jasmine.
   - *Uses*: Relaxation, stress relief, and skincare.

4. *Tree oils*
    - *Examples*: Cedarwood, Tea Tree, Sandalwood, Frankincense.
    - *Uses*: Antimicrobial properties, grounding effects, and skin healing.
5. *Spice oils*
    - *Examples*: Clove, Cinnamon, Ginger, Black Pepper.
    - *Uses*: Warming effects, digestive aid, and immune support.
6. *Mint oils*
    - *Examples*: Peppermint, Spearmint.
    - *Uses*: Energizing, digestive support, and headache relief.
7. *Resin oils*
    - *Examples*: Myrrh, Frankincense.
    - *Uses*: Meditation, spiritual practices, and skin rejuvenation.

8.2 Common uses of essential oils

- *Aromatherapy*: Inhalation of oils to promote emotional and physical well-being.
- *Topical application*: Diluted oils applied to the skin for various benefits.
- *Household cleaning*: Natural disinfectants and deodorizers.
- *Personal care products*: Added to lotions, shampoos, and soaps for fragrance and benefits.
- *Massage*: Used with carrier oils for relaxation and muscle relief.

8.3 Safety and precautions

- *Dilution*: Always dilute essential oils before applying to the skin to prevent irritation.
- *Patch test*: Conduct a patch test to check for allergic reactions.
- *Consultation*: Seek advice from a healthcare provider, especially for pregnant women, children, or those with medical conditions.

8.4 Glucosinolates

Glucosinolates are a group of sulfur-containing compounds found primarily in cruciferous vegetables, such as broccoli, kale, cabbage, and Brussels sprouts. They are known for their potential health benefits, including anti-cancer properties, due to their ability to produce bioactive compounds when broken down.

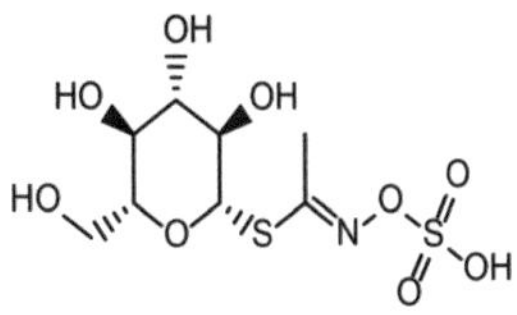

Methyl glucosinolates

When these vegetables are chopped, chewed, or cooked, glucosinolates can be converted into isothiocyanates, indoles, and other compounds that may help reduce the risk of certain cancers and have anti-inflammatory effects. Additionally, they may play a role in detoxification processes in the body.

Including a variety of cruciferous vegetables in your diet can provide these beneficial compounds along with vitamins, minerals, and fiber.

Acknowledgements

The author acknowledges the use of Chat Gpt for language polishing of the manuscript.

Author details

Bibi Rohida
Department of Organic Chemistry, Quaid-i-Azam University Islamabad, Killi Gulzar Western By Pass Quetta, Baluchistan, Pakistan

*Address all correspondence to: rohida@chem.qau.edu.pk

References

[1] Salam U, Ullah S, Tang ZH, Elateeq AA, Khan Y, Khan J, et al. Plant metabolomics: An overview of the role of primary and secondary metabolites against different environmental stress factors. Life. 2023;**13**(3):706

[2] BYJUS. What are Metabolites?—Primary and Secondary Metabolites [Internet]. BYJU'S; 2019. Available from: https://byjus.com/biology/metabolites/

[3] Wikipedia. Secondary Metabolite [Internet]. Wikimedia Foundation; 2019. Available from: https://en.wikipedia.org/wiki/Secondary_metabolite

[4] Tiwari R, Rana CS. Plant secondary metabolites: A review. International Journal of Engineering Research and General Science. 2015;**3**(5):661-670

[5] Atanasov AG, Waltenberger B, Pferschy-Wenzig EM, Linder T, Wawrosch C, Uhrin P, et al. Discovery and resupply of pharmacologically active plant-derived natural products: A review. Biotechnology Advances. 2015;**33**(8):1582-1614

[6] Wani AK, Akhtar N, Sharma A, El-Zahaby SA. Fighting carcinogenesis with plant metabolites by weakening proliferative signaling and disabling replicative immortality networks of rapidly dividing and invading cancerous cells. Current Drug Delivery. 2023;**20**(4):371-386

[7] Elshafie HS, Camele I, Mohamed AA. A comprehensive review on the biological, agricultural and pharmaceutical properties of secondary metabolites based-plant origin. International Journal of Molecular Sciences. 2023;**24**(4):2-20

[8] Marchev AS, Yordanova ZP, Georgiev MI. Green (cell) factories for advanced production of plant secondary metabolites. Critical Reviews in Biotechnology. 2020;**40**(4):443-458

[9] Daba A, Tadesse M, Habte G, Negawo AT, Berecha G. Phytochemical composition of essential oils from aromatic plants inherited with bioherbicidal activity in arabica coffee production system of Ethiopia. Journal of Agriculture and Food Research. 2022;**10**:100368

[10] Liu T, Luo T, Guo X, Zou X, Zhou D, Afrin S, et al. PgMYB2, a MeJA-responsive transcription factor, positively regulates the dammarenediol synthase gene expression in *Panax ginseng*. International Journal of Molecular Sciences. 2019;**20**(9):2219

[11] Pang Z, Chen J, Wang T, Gao C, Li Z, Guo L, et al. Linking plant secondary metabolites and plant microbiomes: A review. Frontiers in Plant Science. 2021;**12**:621276

[12] Singh A. Secondary Metabolites: Definition, Classes, and Functions. Plant Cell Technology; 2022. Available from: https://plantcelltechnology.com/blogs/blog/blog-secondary-metabolites-definition-classes-and-functions part1?srsltid=AfmBOopzOcePVIFg1b0PYCQUhCwBtIzwcklloF_Iwh5Qeo4BnB0jNuNv

[13] Markham KR. Techniques of Flavonoid Identification. London: Academic Press INC; 1982

[14] Naikoo MI, Dar MI, Raghib F, Jaleel H, Ahmad B, Raina A, et al. Role and regulation of plants phenolics in abiotic stress tolerance: An overview. In: Plant Signaling

Molecules. 1st ed, Chapter 9. Elsevier; 2019. pp. 157-168. DOI: 10.1016/B978-0-12-816451-8.00009-5

[15] Karak P. Biological activities of flavonoids: An overview. International Journal of Pharmaceutical Sciences and Research. 2019;**10**(4):1567-1574. Available from: https://byjus.com/biology/metabolites/#:~:text=Metabolites%20are%20intermediate%20end%20products,in%20growth%2C%20reproduction%20and%20development

Chapter 2

Terpenoids: Diverse Structures and Functions in Plant Defense and Communication

Ilham El Arabi, Driss Ousaaid and Hassan Laaroussi

Abstract

This chapter explores the expansive world of terpenoids, one of the largest and most diverse classes of secondary plant metabolites. It begins by introducing the basic isoprene unit and outlining the biosynthetic pathways that lead to the vast array of terpenoid structures found in nature. The chapter then delves into the structural diversity of terpenoids, covering monoterpenes, sesquiterpenes, diterpenes, triterpenes, and other subclasses, highlighting their unique chemical properties. The text further examines the multifaceted roles of terpenoids in plant defense mechanisms, including their function as toxins, repellents, and antimicrobial agents. It also discusses their importance in plant-communication and plant-insect interactions, such as attracting pollinators or natural predators of herbivores. Case studies of well-known terpenoids like menthol, camphor, and artemisinin are presented to illustrate the ecological and pharmacological significance of these compounds. The chapter concludes by exploring the potential applications of terpenoids in agriculture, medicine, and industry, emphasizing their role in the development of new pharmaceuticals and biopesticides.

Keywords: terpenoids, plant defense, chemical ecology, biosynthesis, biopesticides, secondary metabolites, industrial applications

1. Introduction

Natural resources are never-ending supply of bioactive compounds that are well-known for their unique beneficial properties for plants, animals, and humans. In order to create safer and more effective phytomedication against various ailments, there has been a major change in the study of medicinal plants and their phytochemistry since the discovery of various biological properties of phytocompounds that have been collected and extracted from natural resources [1]. The impact of dietary non-nutrient compounds on health and well-being has garnered a lot of attention during the last decade. These compounds are found in plant-based diets with anticipated health-promoting beneficial or toxic effects when integrated in the body [2]. The phytocompounds or secondary metabolites are produced by plants against different biotic and abiotic stresses. These compounds were classified according to their structure and biosynthesis pathways, including phenolic acids, flavonoids, terpenoids, and so on [3].

One of the largest and most varied classes of phytochemicals found in nature are terpenoids, also referred to as isoprenoids, which are one of the largest and most diverse groups of naturally occurring organic compounds. These metabolites are found across all kingdoms of life, but they are particularly abundant in the plant kingdom, where they play critical roles in growth, defense, and communication. Terpenoids derived from the basic five-carbon isoprene unit (C_5H_8), which can be linked in various configurations to form a wide range of structures, from simple linear chains to complex polycyclic molecules. This diversity has led to the identification of over 40,000 unique terpenoid compounds, highlighting their vast chemical and functional variability [4].

The environmental and economic importance of terpenoids cannot be overstated. In plants, they serve as major constituents of essential oils, pigments, and resins, influencing not only plant physiology but also their interactions with the environment. In humans, terpenoids have found extensive use in the pharmaceutical, agricultural, and fragrance industries. This chapter aims to provide an in-depth review of the biosynthetic pathways, structural diversity, and ecological roles of terpenoids, together with a discussion of their practical applications in various fields. At the heart of terpenoid chemistry lies the isoprenic unit, a simple molecule that serves as the building block of all terpenoid structures. The combination of several isoprenic units through processes such as condensation and cyclization leads to the formation of various classes of terpenoids, each with unique properties and functions [5]. This chapter will look at the enzymatic pathways responsible for terpenoid diversity and explore the evolutionary importance of these pathways in the adaptation of plants to their environment.

2. Biosynthesis of terpenoids

The biosynthesis of terpenoids is a complex and highly regulated process that involves multiple enzymatic steps. There are two primary pathways responsible for terpenoid synthesis in plants: the mevalonate (MVA) pathway and the methylerythritol phosphate (MEP) pathway. These pathways operate in distinct cellular compartments and are responsible for the production of specific types of terpenoids.

2.1 The mevalonate (MVA) pathway

The MVA pathway occurs in the cytosol and is the primary source of sesquiterpenes (C_{15}) and triterpenes (C_{30}) in plants. This pathway starts with acetyl-CoA, which undergoes a series of reactions to form mevalonic acid, a crucial intermediate. Mevalonic acid is then phosphorylated and decarboxylated to produce isopentenyl pyrophosphate (IPP), the active isoprene unit that serves as a precursor for longer terpenoid chains [6]. Recent studies have highlighted the importance of enzyme specificity in the MVA pathway. For example, the enzyme farnesyl diphosphate synthase (FPPS) plays a critical role in determining the chain length of the final terpenoid product, influencing the production of various sesquiterpenes that have ecological and medicinal significance [7]. Understanding the regulation of the MVA pathway has significant implications for biotechnological applications, including the production of high-value terpenoids through metabolic engineering [8].

2.2 The methylerythritol phosphate (MEP) pathway

In contrast to the MVA pathway, the MEP pathway takes place in the plastids and is responsible for the biosynthesis of monoterpenes (C_{10}), diterpenes (C_{20}), and tetraterpenes (C_{40}). This pathway starts with pyruvate and glyceraldehyde-3-phosphate, leading to the formation of 1-deoxy-D-xylulose-5-phosphate (DXP), a key intermediate. Subsequent enzymatic steps convert DXP into IPP and dimethylallyl pyrophosphate (DMAPP), which are then utilized for the synthesis of various terpenoids [9]. The regulation of the MEP pathway is tightly linked to environmental conditions, such as light and temperature, which can affect the production of specific terpenoids [10]. For instance, certain volatile monoterpenes are synthesized in response to herbivory or abiotic stress, highlighting the adaptive significance of the pathway [11]. Advances in our understanding of MEP pathway enzymes, such as geranylgeranyl diphosphate synthase (GGPPS), have facilitated the development of transgenic plants with enhanced terpenoid profiles for agricultural and industrial purposes [12]. **Table 1** provides a comparative overview of the MVA and MEP pathways, highlighting their key characteristics and the implicated enzymes.

2.3 Enzymatic modifications and terpenoid

Once the basic terpenoid structure is formed, a range of enzymatic modifications can further diversify the molecule. These modifications include cyclization, hydroxylation, methylation, and glycosylation, each catalyzed by a specific set of enzymes [13]. Terpene synthases (TPS) are particularly important in this regard, as they catalyze the formation of cyclic structures that define many terpenoids' chemical properties [14]. For example, monoterpene synthases can produce a wide array of cyclic and linear monoterpenes from the same precursor, reflecting the plasticity of terpenoid biosynthesis [15]. The diversity of terpenoid structures resulting from these modifications underlies their ecological roles and potential applications, which will be explored in subsequent sections of this chapter. **Table 2** categorizes the major terpenoid classes, providing examples and their primary sources.

Pathway	Cellular compartment	Final products	Key enzymes	Environmental factors
MVA	Cytosol	Sesquiterpenes, Triterpenes	FPPS, HMGR	Low light, High carbon
MEP	Plastid	Monoterpenes, Diterpenes, Tetraterpenes	DXS, GGPPS	High light, Temperature

Table 1.
Comparative analysis of the MVA and MEP pathways in terpenoid biosynthesis.

Terpenoid class	Isoprene units	Example compounds	Natural sources	Ecological/ Pharmacological role
Monoterpenes	2	Limonene, Menthol	*Citrus*, Mint	Pollinator attraction, Defense
Sesquiterpenes	3	Artemisinin, Farnesol	*Artemisia*, Ferns	Antimicrobial, Antimalarial
Diterpenes	4	Taxol, Ginkgolide	Yew, Ginkgo	Anticancer, Defensive toxins
Triterpenes	6	Saponins, Steroids	Various plants	Membrane stabilization, Defense
Tetraterpenes	8	Lycopene, Beta-carotene	Tomatoes, Carrots	Antioxidants, Pigmentation

Table 2.
Classification of terpenoids with examples and ecological significance.

3. Ecological roles of terpenoids

Terpenoids are a diverse and crucial class of secondary metabolites that significantly impact ecosystems, serving essential roles in plant defense and communication. This overview delves into their ecological functions, illustrating their versatility with examples and supporting references. Their chemical diversity not only aids plant survival but also facilitates vital ecological interactions, underscoring their importance in ecosystem dynamics [16].

3.1 Role in plant defense against herbivores and pathogens

Plants have developed sophisticated defense mechanisms to combat herbivores and pathogens, and terpenoids are at the forefront of these strategies. These compounds serve as critical components of the plant's arsenal, functioning as toxins, repellents, and growth inhibitors to deter or neutralize potential threats [17]. When plants face attacks, they employ terpenoids to disrupt the feeding, development, or survival of herbivores and pathogens, enhancing their resilience and adaptability.

3.1.1 Direct defense mechanisms

The defensive role of terpenoids spans across various functions:

- *Insecticidal action*: Terpenoids act as natural insecticides. For instance, limonene, a monoterpene present in citrus peels, effectively repels a variety of insects [18]. Similarly, diterpenoids like resin acids found in pine needles are lethal to many herbivorous insects, hindering their digestion and development [19]. Robust experimental evidence found that terpenoid-based nanoemulsions exhibited an intriguing larvicidal effect against mosquito larvae in comparison to temephos organophosphate as a standard drug used to control mosquitoes. Interestingly, the produced nanoparticles did not interfere with the zebrafish organs' normal functions [20]. Various studies demonstrated that terpenoids could be included as biodrugs environmentally friendly insecticides to lessen the negative impacts of chemical agents [21–23].

- *Antifungal action*: Sesquiterpenes are particularly effective against pathogenic fungi. A notable example is farnesene, a sesquiterpene produced by species in the Asteraceae family. It inhibits spore germination and impedes fungal growth, preventing infections. These compounds also discourage herbivores, creating a dual defense strategy [24]. The low molecular weight and lipophilic character of terpenoids make them excellent molecules that hinder the sporulation and germination of various fungi causing cell death [25]. Furthermore, they can disrupt the cell membrane by interacting with vesicles and cell membranes thereby impairing ergosterol biosynthesis in Candida strains. This results in osmotic and metabolic instability that affects the integrity of the cell membrane [25, 26]. Another study found that limonene and thymoquinone can destabilize the fungal cell wall by inhibiting the formation of β-glucan and chitin [26].

- *Antimicrobial action*: Beyond insect and fungal defenses, some terpenoids exhibit broad-spectrum antimicrobial properties. Geraniol, found in geranium essential oils, has shown potent activity against bacteria and fungi, protecting the plant from various diseases. This antimicrobial capability underscores the multifunctional nature of terpenoids in plant defense [27]. Wang et al. found that various terpoinds, including α-pinene, limonene, myrecine, geraniol, linalool, nerol, and terpineol exhibited an important antibacterial effect against different [28] *Staphylococcus aureus* with MIC50 and MBC values ranging between 0.420 and 1.598 mg/ml, and 0.673 and 3.432 mg/ml, respectively [29].

3.1.2 Indirect defense mechanisms

In addition to direct deterrence, plants have evolved indirect defense strategies that rely on attracting natural predators of herbivores through the emission of specific terpenes. This form of biological pest control helps in reducing herbivore populations without direct chemical action:

- Maize and caryophyllene emission: When maize is attacked by caterpillars, it releases volatile caryophyllenes. These compounds act as signals that attract parasitoid wasps, which then lay eggs inside the caterpillar, eventually killing it and protecting the maize [28].

- Tomato and ladybug attraction: In the case of aphid infestations, tomato plants emit terpenes that draw ladybugs, natural predators of aphids. This terpene-mediated signaling encourages biological control, showcasing the strategic use of volatile compounds in indirect defense [30].

3.1.3 Mechanism of terpenoid induction

The production of terpenoids is a dynamic and responsive process, often triggered by environmental stimuli such as herbivory or pathogen attack. Damage to plant tissues activates internal signaling pathways, notably the jasmonic acid pathway. Jasmonic acid, a key phytohormone, is released upon injury and stimulates the synthesis of terpenoids. This creates a flexible and adaptive defense system that allows plants to quickly respond to evolving threats (**Table 3**) [31].

This integrated and enriched perspective underscores the multifaceted role of terpenoids in plant defense. Through direct toxicity and indirect signaling, plants harness the chemical complexity of terpenoids to maintain ecological balance and protect themselves from a broad spectrum of natural enemies.

Terpenoid	Source	Ecological function	Reference
Limonene	Citrus peels	Insect repellent	[13]
Artemisinin	*Artemisia annua*	Antimalarial	[16]
Farnesene	Asteraceae species	Antifungal	[14]
Resin acids	Conifer needles	Insect deterrence	[13]
Geraniol	Geranium oils	Antimicrobial	[15]
Beta-phellandrene	Tomato plants	Attraction of ladybugs against aphids	[30]
Caryophyllene	Maize (under attack)	Attraction of parasitoid wasps	[24]

Table 3.
Different terpenoids, their sources, and their ecological functions.

The production of these defensive terpenoids is often induced by environmental triggers, such as herbivory or pathogen attacks, which activate the plant's internal signaling pathways. The release of jasmonic acid following damage is known to stimulate the synthesis of defensive terpenoids, creating a dynamic and responsive defense system [31].

3.2 Ecological roles of volatile terpenoids in plant-insect interactions

Volatile terpenoids play a pivotal role in plant communication, serving both as attractants for beneficial organisms and as deterrents against herbivores. These compounds are key to facilitating pollination and protecting plants from threats, showcasing their dual ecological function in plant-insect interactions. To attract pollinators, many plants release specific terpenoids. For example, lavender emits linalool, a monoterpene, which acts as a potent attractant for bees and other pollinators, ensuring successful reproduction. Similarly, some orchids emit terpenoid compounds that mimic insect pheromones to attract particular pollinators [32]. These interactions illustrate how terpenoids mediate complex ecological relationships, enhancing both pollination efficiency and plant fitness. In addition to attracting pollinators, volatile terpenoids are crucial in plant defense. They can directly deter herbivores by signaling toxicity or unpleasant taste. A well-known example is maize, which produces caryophyllene to deter herbivores, effectively reducing damage [33].

Furthermore, volatile terpenoids contribute to indirect defenses by recruiting natural predators of herbivores. When attacked, some plants release terpenoid-based signals that attract predators, such as parasitoid wasps, which target the herbivores feeding on the plant [34, 35]. This indirect defense strategy, often described as a "cry for help," helps maintain ecological balance. The multifaceted roles of volatile terpenoids underscore their importance in sustaining both reproductive success and protection, highlighting their essential function in plant communication and ecosystem stability. **Table 4** summarizes key examples of terpenoids and their roles in plant defense and attraction, underlining their ecological significance.

3.3 Impact of terpenoids on ecological communities and plant communication

Terpenoids play crucial roles not only in direct plant-insect interactions but also in broader ecological communities and plant communication. These compounds are involved in shaping interactions within ecosystems, mediating behaviors, and influencing the composition of various species.

Compound	Terpenoid class	Ecological role	Target organisms
Alpha-pinene	Monoterpene	Herbivore Deterrent	Insect Pests (Beetles, Moths)
Farnesene	Sesquiterpene	Repellent/Alarm Signal	Aphids, Herbivorous Insects
Linalool	Monoterpene	Pollinator Attraction	Bees, Butterflies
Artemisinin	Sesquiterpene	Antimalarial Defense	Pathogenic Parasites
Taxol	Diterpene	Chemical Defense	Fungal Pathogens, Herbivores

Sources: Pichersky and Gershenzon [35]; Chen et al. [34].

Table 4.
Examples of terpenoids involved in defense and attraction.

In boreal forests, conifer trees like pines emit substantial amounts of monoterpenes, such as pinene, which serve as defense mechanisms against insects and inhibit microbial competitors. Additionally, these emissions affect local atmospheric chemistry, contributing to cloud formation and impacting climate patterns [36]. In Mediterranean aromatic plant habitats, plants like rosemary and sage release terpenes to deter herbivores while attracting specific pollinators, thus influencing local insect communities by altering population compositions [37].

Terpenoids are also integral to chemical signaling in plants, contributing to various ecological interactions. For example, many flowering plants emit fragrant terpenoids like linalool to attract pollinators such as bees and butterflies, facilitating pollination [38]. The specificity of these terpenoid-based scents ensures effective pollinator-plant relationships. In addition, some terpenoids exhibit allelopathic properties, allowing plants to affect the growth and development of neighboring species. This can involve inhibiting seed germination, root elongation, or nutrient uptake in competing plants, thereby conferring a competitive advantage to the producing plant [39].

4. Applications of terpenoids in agriculture, medicine, and industry

Due to their diverse chemical properties, terpenoids have found numerous applications in agriculture, medicine, and industry.

The ecological roles of terpenoids are closely tied to their practical applications in agriculture, medicine, and various industries. This section explores how terpenoids have been harnessed to benefit human activities, from pest control to pharmaceutical development.

4.1 Agricultural uses: Biopesticides and plant growth regulators

Due to the increased awareness of environmental sustainability, terpenoids are being recognized as valuable alternatives to synthetic agrochemicals. They offer effective pest control while minimizing ecological impact.

Biopesticides: Recent research has underscored the efficacy of terpenoid-based biopesticides, such as pyrethrins, which are derived from Chrysanthemum flowers. These compounds are powerful insecticides with low toxicity to mammals, making them suitable for integrated pest management (IPM) systems. A review published in plants has shown that these natural insecticides are effective against a broad spectrum of agricultural pests while posing minimal risk to non-target species [40]. Additionally, monoterpenoids like thymol, found in essential oils, have demonstrated significant antifungal and insect-repellent properties, making them promising candidates for sustainable agricultural practices [41].

Plant Growth Regulation: Terpenoids are not only used for pest control but also play a crucial role in plant growth. Diterpenoids such as gibberellins are essential for regulating various developmental processes, including stem elongation, seed germination, and fruit development. Recent studies have emphasized their use in agriculture to enhance crop yield, regulate flowering time, and improve fruit quality. In particular, advancements in biotechnology have allowed for more efficient production of gibberellins, reducing costs and increasing accessibility for agricultural applications [42].

These applications underscore the importance of biologically-based solutions in modern agriculture, aligning with global goals for sustainability and reduced chemical dependency.

4.2 Pharmaceutical applications: Focus on malaria and cancer treatments

Terpenoids are a cornerstone in pharmaceutical innovation due to their diverse bioactive properties, offering effective treatment options for various diseases.

4.2.1 Antimalarial drugs

Over the past two decades, natural products have been extensively studied and identified as promising lead compounds in the development of drugs to combat various protozoan diseases, including malaria.

Artemisinin, a sesquiterpene lactone derived from *Artemisia annua*, remains crucial in combating malaria. It has shown rapid efficacy against *Plasmodium* parasites, which cause malaria, making it a first-line treatment in many countries. Recent research focuses on developing semi-synthetic derivatives of artemisinin to enhance stability and efficacy, given the challenges of natural artemisinin extraction. A study in The Lancet Infectious Diseases highlights ongoing advances in artemisinin combination therapies to counteract emerging drug resistance [33].

4.2.2 Antitumor and anticancer effects: Insights into molecular mechanisms

Terpenoids, a diverse class of natural compounds derived from plants, have garnered significant attention for their potential therapeutic properties, particularly in cancer research. These compounds, including monoterpenes, sesquiterpenes, diterpenes, and triterpenes are well known for their immunomodulatory, anti-inflammatory, and cytotoxic effects [43]. Nowadays, several studies have highlighted their ability to induce apoptosis, inhibit metastasis, prevent tumor angiogenesis, and thus to modulate cancer cell growth.

It has been documented that thymol or 2-isopropyl-5-methyl phenol, a natural monoterpenoid phenolic compound found in various plants belonging to the Lamiaceae family such as s thyme (*Thymus vulgaris*), rosemary (*Rosmarinus officinalis*), and oregano (*Origanum vulgare*) stops the cell cycle at G0/G1 phase in Promyelocytic Leukemia Cells (HL-60), preventing its progression into the S phase and thereby inhibiting cell proliferation, which leads to its anticancer effect. Furthermore, it induces mitochondrial depolarization, leading to ATP depletion and increased pro-apoptotic factors [44]. Accordingly, thymol down-regulates the Bcl-2 gene expression while up-regulating Bax protein expression, thereby triggering the activation of caspases and promoting the release of cytochrome C, Omi/HtrA2 (serine protease), and Smac/DIABLO (second mitochondria-derived activator of caspases) as key pro-apoptotic factors. Likewise, Altintas and coworkers have reported that thymol exerts its anticancer effect on the hepatocellular carcinoma (HepG2) cell line by reducing oxidative stress, inducing DNA damage and activating the apoptotic pathways [45]. More in-deep investigations highlighted that thymol can also interact with other signaling pathways that involve calcium ion (Ca^{2+}) signaling. In fact, it increases the release of Ca^{2+} from the endoplasmic reticulum (ER) through a phospholipase C (PLC)-dependent mechanism. As an inositol 1,4,5-trisphosphate (IP_3)-mediated release signaling pathway, the increased intracellular Ca^{2+} concentration is followed by an influx of additional Ca^{2+} ions from the extracellular space, leading to high intracellular Ca^{2+} levels. This molecular signaling pathway has been shown to promote apoptosis, or programmed cell death, in various cancer cell types, including MG63 human osteosarcoma cells, DBTRG-05MG human glioblastoma cells, and PC3 human prostate cancer cells [46–48]. Furthermore, menthol, another valuable monoterpene compound has been reported to be effective against multiple human cancers. In this framework, Li and coworkers investigated the beneficial effect of menthol on SNU-5 human gastric cancer cells [49]. Obtained data showed that menthol treatment, as its anticancer property, lead to a reduction in the activity of topoisomerase I and II, resulting in DNA damage and subsequent SNU-5 gastric cell death. Eventually, targeting this signaling way could be a promising therapeutic strategy for cancer management, especially if it is combined with other clinical approaches.

In addition to monoterpenes $(C_5H_8)_2$, sesquiterpenes $(C_{15}H_{24})$ have also been investigated for their potential anticancer effect. For instance, compounds like dindol-01, coronopilin, and ambrosin have been reported to exert DNA damage in breast cancer cell lines (MCF-10A, MCF-7, JIMT-1 and HCC1937) as well as arrest the cell cycle at the S and G2 phases, enhancing p53 protein expression as a main tumor-suppressor mechanism in cancer cells [50]. Moreover, a natural Sesquiterpene lactone (SQL) extracted from *Artemisia macrocephala* exhibited *in vitro* its anticancer action against MCF-7, 3 T3, and HeLa cells; and mild to moderate antinociceptive activity through Reactive Oxygen species (ROS) generation. Additionally, SQL disturbs the cells calcium homeostasis by inhibiting the *Sarcoplasmic Reticulum Calcium ATPase (SERCA) pump*. It also down-regulates the NF-κB signaling pathway, reducing inflammation, cancer cell survival, and metastasis while limiting angiogenesis, thereby starving tumors of necessary nutrients. In contrast, it up-regulates the p53 gene expression, enhancing tumor suppression by promoting cell cycle arrest and apoptosis [51].

Taxol (paclitaxel), a diterpenoid $(C_{20}H_{32})$ extracted from the Pacific yew tree (*Taxus brevifolia*), continues to be a leading chemotherapeutic agent for various cancers, including breast and ovarian cancer. Recent publications in Nature Reviews Drug Discovery emphasize the development of synthetic biology approaches to produce taxol via microbial fermentation, aiming to address supply constraints and environmental concerns associated with natural extraction [34]. These methods are improving yield and reducing the ecological footprint of taxol production, making it more sustainable.

Triterpene compounds $(C_{30}H_{48})$ have also shown promising potential in cancer treatment due to their diverse chemical structure, targeting multiple signaling pathways with specific mechanisms of action. Among these molecules, oleanolic acid, ursolic acid, and betulinic acid have been extensively examined. For instance, oleanolic acid and ursolic, which have nearly identical chemical structures, have been found to exhibit antitumoral effect by inducing apoptosis in human cancer cell lines (HepG2, Hep3B, Huh7 and HA22T) [52]. Accordingly, both compounds both compounds appear to act through similar mechanisms, mainly by decreasing mitochondrial membrane potential and Na^+-K^+-ATPase activity; increasing DNA damage and caspase-3 and caspase-8 activities. Furthermore, they reduce the production of VEGF and ICAM-1 and suppress cancer cell adhesion.

These examples highlight the potential of terpenoids as safer and promising candidates for the development of new effective anticancer drugs.

4.3 Clinical trials and patent developments of terpenoids as potential anticancer agents

Clinical trials are a crucial step in the drug discovery process, serving as a link between experimental research and the approval of new treatments for clinical use, as they provide essential data on the safety, efficacy, and recommendations for optimal use, targeting the safer application for the general population.

As documented in **Table 5**, bioactive molecules belonging to terpenoids family have been the subjects of numerous clinical trial studies. It is most relevant to indicate that the present data is incomplete, and in most clinical investigations the results are not available or are still pending. In general, the reported results highlight some clinical trials investigating the use of specific terpenoid molecules in various cancer treatment with a focus on doses, administration routes, treatment durations, sex, and inclusion criteria. For instance, *artesunate*, a sesquiterpene lactone derived from artemisinin, has been investigated against solid tumor and lung cancer in a phase I clinical trial (NCT02353026 and NCT05166616, respectively). Obtained results indicated that the intravenous *artesunate treatment* (NCT02353026) *generated different depending dose-effects. Most importantly,* 18 mg/kg was selected as a maximum tolerated dose (MTD) of intravenous artesunate on a day 1/day 8, 3-week cycle (**Table 5**). Similarly, at the selected dose, artesunate exerts modest clinical activity. However, the observed side effects were similar to those reported in malaria treatment reports, including anemia, nausea, vomiting, fatigue, anorexia, electrolyte disturbances, and liver issues [53]. Moreover, a single-center, randomized, double-blind, placebo-controlled trial reported that the oral artesunate treatment (200 mg) for 14 consecutive days exhibited remarkable antiproliferative effects in colorectal cancer by increasing the Cluster of Differentiation 31(CD31) expression and reducing the Ki67 [54]. Retinol, a monoterpenoid bio-compound derived from beta-carotene, also known as Vitamin A has also been the subject of the phase I clinical trial (NCT03870529) to assess its potential therapeutic effect as an adjuvant anticancer therapy (**Table 5**). Accordingly, following a 21-day treatment period, the trial demonstrated that retinol increases the number of germinal centers (immune structures that help mature antibodies) in tumor and lymphatic tissues, which may be beneficial for cancer patients. These findings suggest that retinol could be beneficial as an adjunctive therapy in cancer treatment, particularly in boosting the immune system's ability to respond to tumor cells.

Menthol and triptolide have been investigated for breast cancer (NCT01855607) and adenosquamous carcinoma of the pancreas (NCT04896073), respectively as part of phase II clinical trial. Obtained results indicated that the topical application of menthol (1%) significantly mitigates breast cancer symptoms.

The use of terpenoids as active supplements or/and adjuvant treatments (Phase III clinical trials) has recently been adopted as a useful management strategy for stomach (NCT01411189), gastric (NCT01411176), cervical (NCT02255084), and colorectal (NCT02588248) cancers, as well as their associated health issues (**Table 5**). The common denominator among these clinical trials is the incorporation of terpenoids as functional micronutrients either in the prevention or treatment of cancer in virtue of their antioxidant, anti-inflammatory and immunomodulatory properties [55–57]. In fact, the multi-targeted mechanisms of action of discussed biomolecules make them promising candidates for cancer prevention and treatment, with ongoing research focusing on their ability to complement and boost conventional cancer therapies.

Nowadays different patents have been registered across various countries, particularly, China (Patents No 1020200023123, 108652091, 104,327,152, 103,393,598, 108,025,070), Japan (WO/2017/131175, WO/2017/043613), India (201,941,035,781, 201,827,013,711, 201,941,045,086), United States of America (20,180,326,052, 20,190,343,909), Russian Federation (02233661), and Singapore (11,201,802,915 W). These patents highlight the growing interest in developing potential anticancer agents based on terpene molecules such as thymol, menthol, D-limonene sesquiterpene lactones, elemene, triptolide, and phytol.

Terpenoids	Clinical trial study	Simple size/ Age/ Sex	NCT number	Phase	Status
Menthol	Breast cancer	60/21 year and older/ females	NCT01855607	II	Unknown
	Stomach cancer	33/20 years and older/ both sexes	NCT01411189	III	Completed
	Gastric cancer	85/adults patients/ both sexes	NCT01411176	III	Completed
	Cervical cancer	1806/30–65 years/ females	NCT02255084	III	Completed
	colorectal cancer	37/45 years and older/ both sexes	NCT02588248	III	Unknown
Artesunate	Solid tumor	19/both sexes	NCT02353026	I	Completed
	Colorectal cancer	200/Adults/ both sexes	NCT03093129	II	Completed
	Breast cancer	23/Adults/ Females	NCT00764036	I	Completed
	Bowel and colorectal cancer	200/both sexes/	NCT02633098	II	Ongoing
Triptolide	Lung Cancer	18 Years and older /both sexes	NCT05166616	I	Ongoing
	Colorectal, Pancreatic Gastric, Breast, Prostate, Cancers, Solid Carcinoma, and Solid Tumors	600/adults/ both sexes	NCT03129139	I	Recruiting
	Adenosquamous carcinoma of pancreas	55/adults/ both sexes	NCT04896073	II	Ongoing
Retinol	Lung cancer	20/18 Years and older/ both sexes	NCT03870529	I	completed
	Head and Neck Cancer	30/18 Years and older	NCT06100692	—	Not yet recruiting
	Lung cancer	18,314/ 45 to 69 years/ both sexes	NCT00712647		Completed
Thymoquinone	Oral malignant tumor	48/ 18 Years to 75 Years/ both sexes	NCT03208790	II	Completed
	Polycystic ovary syndrome	18 Years to 35 Years/ Females	NCT04852510	—	Completed

Terpenoids	Clinical trial study	Simple size/ Age/ Sex	NCT number	Phase	Status
Limonene	Breast cancer	18 Years and older/ Females	NCT01046929	I	Completed
	Breast cancer	18 Years to 65 Years/ Females	NCT01459172	I	Complted
Elemene	Liver cancer	18 to 75 Years/ both sexes	NCT03166553	—	Unknown
	Lung cancer	80/adults/ both sexes	NCT03123484	II	Unknown
	Pancreatic cancer	19	NCT03117920	II	Completed
	Non-small cell lung cancer	18 Years to 75 Years/ both sexes	NCT03123484	II	Unknown

Table 5.
Terpenoids as adjuvant therapy in the management of cancer: Clinical trials.

All these patents highlight the potential role of different terpenoid molecules as potential anticancer agents for different types of cancer, reflecting significant international research and development in the field.

4.4 Industrial applications: Fragrances, flavors, and biofuels

Terpenoids play a pivotal role in various industries due to their aromatic properties and potential as renewable energy sources.

Essential oils and perfumery: Terpenoids like limonene, menthol, and pinene are widely utilized in the fragrance and flavor industries due to their pleasant aromas and versatility. In recent years, consumer demand for natural and sustainable ingredients has driven the expansion of essential oils in perfumery and cosmetics. A recent review in Industrial Crops and Products discusses the role of terpenoids in developing eco-friendly fragrances and the shift toward organic compounds over synthetic alternatives [58].

Biofuels: The search for renewable energy has turned to terpenoids as potential candidates for biofuel production due to their high energy density. Advances in genetic engineering and synthetic biology have enabled the microbial production of terpenoid-based biofuels, reducing dependency on fossil fuels. Research published in Energy and Environmental Science explores the genetic modifications in microorganisms to increase terpenoid yield, focusing on optimizing the cost-effectiveness of biofuel production [59]. This research highlights the role of terpenoids in the development of sustainable energy solutions.

These applications underscore the industrial relevance of terpenoids, promoting both sustainability and innovation across various sectors.

5. Conclusion

Terpenoids represent a vast and versatile class of natural compounds with significant ecological, pharmacological, and industrial relevance. Their roles in plant defense and communication illustrate the complexity of plant interactions with their environment, while their diverse chemical structures provide a rich source of compounds for human use. As research continues to uncover the molecular mechanisms underlying terpenoid biosynthesis and function, there is growing potential for the development of new applications in medicine, agriculture, and beyond.

This chapter has provided an overview of the current state of knowledge on terpenoids, with a focus on their biosynthesis, ecological roles, and practical applications. Future studies are expected to further unravel the complexities of terpenoid metabolism and open new avenues for biotechnological innovation.

Conflict of interest

The authors declare no conflict of interest.

Author details

Ilham El Arabi*, Driss Ousaaid and Hassan Laaroussi
University Sidi Mohammed Ben Abdellah, Fez, Morocco

*Address all correspondence to: ilham.elarabi@gmail.com
and ilham.elarabi@usmba.ac.ma

References

[1] Clark AM. Natural products as a resource for new drugs. Pharmaceutical Research. 1996;**13**(8):1133-1141. DOI: 10.1023/A:1016091631721

[2] Gry J, Black L, Eriksen FD, Pilegaard K, Plumb J, Rhodes M, et al. EuroFIR-BASIS – A combined composition and biological activity database for bioactive compounds in plant-based foods. Trends in Food Science and Technology. 2007;**18**(8):434-444. DOI: 10.1016/j.tifs.2007.05.008

[3] Karasawa MMG, Mohan C. Fruits as prospective reserves of bioactive compounds: A review. Natural Products and Bioprospecting. 2018;**8**:335-346. DOI: 10.1007/s13659-018-0186-6

[4] Bergman ME, Dudareva N. Plant specialized metabolism: Diversity of terpene synthases and their products. Current Opinion in Plant Biology. 2024;**81**:102607. DOI: 10.1016/j.pbi.2024.102607

[5] Gershenzon J, Dudareva N. The function of terpene natural products in the natural world. Nature Chemical Biology. 2007;**3**(7):408-414. DOI: 10.1038/nchembio.2007.5

[6] Vranová E, Coman D, Gruissem W. Network analysis of the MVA and MEP pathways for isoprenoid synthesis. Annual Review of Plant Biology. 2013;**64**(1):665-700. DOI: 10.1146/annurev-arplant-050312-120116

[7] Richter A, Seidl-Adams I, Köllner TG, Schaff C, Tumlinson JH, Degenhardt J. A small, differentially regulated family of farnesyl diphosphate synthases in maize (*Zea mays*) provides farnesyl diphosphate for the biosynthesis of herbivore-induced sesquiterpenes. Planta. 2015;**241**(6):1351-1361. DOI: 10.1007/s00425-015-2254-z

[8] Kirby J, Keasling JD. Biosynthesis of plant isoprenoids: Perspectives for microbial engineering. Annual Review of Plant Biology. 2009;**60**(1):335-355. DOI: 10.1146/annurev.arplant.043008.091955

[9] Rodrıguez-Concepción M, Boronat A. Elucidation of the methylerythritol phosphate pathway for isoprenoid biosynthesis in bacteria and plastids. A metabolic milestone achieved through genomics. Plant Physiology. 2002;**130**(3):1079-1089. DOI: 10.1104/pp.007138

[10] Banerjee A, Sharkey TD. Methylerythritol 4-phosphate (MEP) pathway metabolic regulation. Natural Product Reports. 2014;**31**(8):1043-1055. DOI: 10.1039/C3NP70124G

[11] Dudareva N, Pichersky E, Gershenzon J. Biochemistry of plant volatiles. Plant Physiology. 2004;**135**(4):1893-1902. DOI: 10.1104/pp.104.049981

[12] Lange BM. The evolution of plant secretory structures and emergence of terpenoid chemical diversity. Annual Review of Plant Biology. 2015;**66**(1):139-159. DOI: doi.org/10.1146/annurev-arplant-043014-114639

[13] Huang H, Liang J, Tan Q, Ou L, Li X, Zhong C, et al. Insights into triterpene synthesis and unsaturated fatty-acid accumulation provided by chromosomal-level genome analysis of Akebia trifoliata subsp. australis. Horticulture Research. 2021;**8**(1):1-15. DOI: doi.org/10.1038/s41438-020-00458-y

[14] Christianson DW. Structural biology and chemistry of the

terpenoid cyclases. Chemical Reviews. 2006;**106**(8):3412-3442

[15] Jin J, Panicker D, Wang Q, Kim MJ, Liu J, Yin J-L, et al. Next generation sequencing unravels the biosynthetic ability of spearmint (Mentha spicata) peltate glandular trichomes through comparative transcriptomics. BMC Plant Biology. 2014;**14**(1):292. DOI: 10.1186/s12870-014-0292-5

[16] Mishra M, Lomate PR, Joshi RS, Punekar SA, Gupta VS, Giri AP. Ecological turmoil in evolutionary dynamics of plant–insect interactions: Defense to offence. Planta. 2015;**242**(4):761-771

[17] Harborne JB. Introduction to Ecological Biochemistry. 1982. Available from: https://www.cabidigitallibrary.org/doi/full/10.5555/19820310798

[18] Sharma S, Loach N, Gupta S, Mohan L. Evaluation of larval toxicity, mode of action and chemical composition of citrus essential oils against Anopheles stephensi and Culex quinquefasciatus. Biocatalysis and Agricultural Biotechnology. 2022;**39**:102284

[19] Alicandri E, Covino S, Sebastiani B, Paolacci AR, Badiani M, Manti F, et al. Diterpene resin acids and olefins in Calabrian pine (Pinus nigra subsp. laricio (Poiret) Maire) oleoresin: GC-MS profiling of major diterpenoids in different plant organs, molecular identification and expression analysis of diterpene synthase genes. Plants. 2021;**10**(11):2391

[20] Duarte JL, Duchon S, Filippo LDD, Chorilli M, Corbel V. Larvicidal properties of terpenoid-based nanoemulsions against the dengue vector *Aedes aegypti L.* and their potential toxicity against non-target organism. PLoS One. 2024;**19**(2):e0293124

[21] Saleem U, Asrar M, Jabeen F, Hussain SM, Hussain D. Determination of insecticidal potential of selected plant extracts against fall armyworm (*Spodoptera frugiperda*) larvae. Heliyon. 2024;**10**(20):e39593. DOI: 10.1016/j.heliyon.2024.e39593

[22] Chaudhary S, Yadav SK, Verma P, Sagar S, Lal M, Chaudhary S, et al. Botanical insecticides for crop protection: Major classes and possible mechanisms of action. In: Insecticides in Pest Control—Impact, Challenges and Strategies. London, UK: IntechOpen; 2024. DOI: 10.5772/intechopen.1006743

[23] Pradana WI, Setiadi EC, Ismail D, Utami TS, Arbianti R, Dadang D, et al. Botanical insecticide production for mealybug (*Dysmicoccus neobrevipes*) pest control using pineapple waste mixture through ultrasonic-assisted extraction method by adjusting type of solvents. AIP Conference Proceedings. 2024;**3215**(1):070019

[24] Zhou F, Pichersky E. More is better: The diversity of terpene metabolism in plants. Current Opinion in Plant Biology. 2020;**55**:1-10

[25] Câmara JS, Perestrelo R, Ferreira R, Berenguer CV, Pereira JAM, Castilho PC. Plant-derived terpenoids: A plethora of bioactive compounds with several health functions and industrial applications—A comprehensive overview. Molecules. 2024;**29**(16):3861

[26] İşcan G, İşcan A, Demirci F. Anticandidal effects of thymoquinone: Mode of action determined by transmission electron microscopy (TEM). Natural Product Communications. 2016;**11**(7):1934578X1601100726

[27] Mabou FD, Yossa IBN. TERPENES: Structural classification and biological

activities. IOSR Journal of Pharmacy and Biological Sciences. 2021;**16**:25-40

[28] Kutty NN, Mishra M. Dynamic distress calls: Volatile info chemicals induce and regulate defense responses during herbivory. Frontiers in Plant Science. 2023;**14**:1135000

[29] Wang C-Y, Chen Y-W, Hou C-Y. Antioxidant and antibacterial activity of seven predominant terpenoids. International Journal of Food Properties. 2019;**22**(1):230-238

[30] Schaller A. Induced Plant Resistance to Herbivory. Dordrecht: Springer Netherlands; 2008. pp. 2340-2354. DOI: 10.1007/978-1-4020-8182-8

[31] Dötterl S, Gershenzon J. Chemistry, biosynthesis and biology of floral volatiles: Roles in pollination and other functions. Natural Product Reports. 2023;**40**(12):1901-1937

[32] Lunau K, Wester P. Mimicry and deception in pollination. Advances in Botanical Research (Elsevier, Academic Press). 2017;**82**:259-279

[33] Montejano-Ramírez V, Ávila-Oviedo JL, Campos-Mendoza FJ, Valencia-Cantero E. Microbial volatile organic compounds: Insights into plant defense. Plants. 2024;**13**(15):2013

[34] Chen F, Tholl D, Bohlmann J, Pichersky E. The family of terpene synthases in plants: A mid-size family of genes for specialized metabolism that is highly diversified throughout the kingdom. The Plant Journal. 2011;**66**(1):212-229

[35] Pichersky E, Gershenzon J. The formation and function of plant volatiles: Perfumes for pollinator attraction and defense. Current Opinion in Plant Biology. 2002;**5**(3):237-243

[36] Adamczyk S, Adamczyk B, Kitunen V, Smolander A. Monoterpenes and higher terpenes may inhibit enzyme activities in boreal forest soil. Soil Biology and Biochemistry. 2015;**87**:59-66

[37] Piccolella S, Crescente G, Pacifico F, Pacifico S. Wild aromatic plants bioactivity: A function of their (poly)phenol seasonality? A case study from Mediterranean area. Phytochemistry Reviews. 2018;**17**(4):785-799

[38] Dudareva N, Pichersky E. Biochemical and molecular genetic aspects of floral scents. Plant Physiology. 2000;**122**(3):627-634

[39] Not Available NA, Duke SO. Ecophysiological aspects of allelopathy. Planta. 2003;**217**(4):529-539

[40] Smith GH, Roberts JM, Pope TW. Terpene based biopesticides as potential alternatives to synthetic insecticides for control of aphid pests on protected ornamentals. Crop Protection. 2018;**110**:125-130

[41] Hezakiel HE, Thampi M, Rebello S, Sheikhmoideen JM. Biopesticides: A green approach towards agricultural pests. Applied Biochemistry and Biotechnology. 2024;**196**(8):5533-5562

[42] Hernández Rodríguez A, Díaz Pacheco A, Martínez Tolibia SE, Melendez Xicohtencatl Y, Granados Balbuena SY, López y López VE. Bioprocess of gibberellic acid by fusarium fujikuroi: The challenge of regulation, raw materials, and product yields. Journal of Fungi. 2024;**10**(6):418

[43] Deb DD, Parimala G, Devi SS, Chakraborty T. Effect of thymol on peripheral blood mononuclear cell PBMC and acute promyelotic cancer cell line HL-60. Chemico-Biological Interactions. 2011;**193**(1):97-106

[44] Yang W, Chen X, Li Y, Guo S, Wang Z, Yu X. Advances in pharmacological activities of terpenoids. Natural Product Communications. 2020;**15**(3):1934578X20903555

[45] Altintas F, Tunc-Ata M, Secme M, Kucukatay V. The anticancer effects of thymol on HepG2 cell line. Medical Oncology. 2023;**40**(9):260

[46] Hsu S-S, Lin K-L, Chou C-T, Chiang A-J, Liang W-Z, Chang H-T, et al. Effect of thymol on Ca2+ homeostasis and viability in human glioblastoma cells. European Journal of Pharmacology. 2011;**670**(1):85-91

[47] Yeh J-H, Chou C-T, Chen I-S, Lu T, Lin K-L, Yu C-C, et al. Effect of thymol on Ca^(2+) homeostasis and viability in PC3 human prostate cancer cells. The Chinese Journal of Physiology. 2017;**60**(1):32-40

[48] Chang H-T, Hsu S-S, Chou C-T, Cheng J-S, Wang J-L, Lin K-L, et al. Effect of thymol on Ca2+ homeostasis and viability in MG63 human osteosarcoma cells. Pharmacology. 2011;**88**(3-4):201-212

[49] Lu H-F, Liu J-Y, Hsueh S-C, Yang Y-Y, Yang J-S, Tan T-W, et al. (−)-menthol inhibits WEHI-3 leukemia cells in vitro and in vivo. In Vivo. 2007;**21**(2):285-289

[50] Sotillo WS, Villagomez R, Smiljanic S, Huang X, Malakpour A, Kempengren S, et al. Anti-cancer stem cell activity of a sesquiterpene lactone isolated from Ambrosia arborescens and of a synthetic derivative. PLoS One. 2017;**12**(9):e0184304

[51] Shoaib M, Shah I, Ali N, Adhikari A, Tahir MN, Shah SWA, et al. Sesquiterpene lactone! A promising antioxidant, anticancer and moderate antinociceptive agent from artemisia macrocephala jacquem. BMC Complementary and Alternative Medicine. 2017;**17**(1):27

[52] Yan S, Huang C, Wu S, Yin M. Oleanolic acid and ursolic acid induce apoptosis in four human liver cancer cell lines. Toxicology In Vitro. 2010;**24**(3):842-848

[53] Deeken JF, Wang H, Hartley M, Cheema AK, Smaglo B, Hwang JJ, et al. A phase I study of intravenous artesunate in patients with advanced solid tumor malignancies. Cancer Chemotherapy and Pharmacology. 2018;**81**(3):587-596

[54] de Vasconcelos Cerqueira Braz J, Carvalho Nascimento Juúnior JA, Serafini MR. Terpenes with antitumor activity: A patent review. Recent Patents on Anti-Cancer Drug Discovery. 2020;**15**(4):321-328

[55] Spina R, Ropars A, Bouazzi S, Dadi S, Lemiere P, Dupire F, et al. Screening of anti-inflammatory activity and metabolomics analysis of endophytic fungal extracts; identification and characterization of perylenequinones and terpenoids from the interesting active alternaria endophyte. Molecules. 2023;**28**(18):6531

[56] Zhou X, Zeng M, Huang F, Qin G, Song Z, Liu F. The potential role of plant secondary metabolites on antifungal and immunomodulatory effect. Applied Microbiology and Biotechnology. 2023;**107**(14):4471-4492

[57] Gutiérrez-del-Río I, López-Ibáñez S, Magadán-Corpas P, Fernández-Calleja L, Pérez-Valero Á, Tuñón-Granda M, et al. Terpenoids and polyphenols as natural antioxidant agents in food preservation. Antioxidants. 2021;**10**(8):1264

[58] Pavela R. Essential oils for the development of eco-friendly mosquito larvicides: A review. Industrial Crops and Products. 2015;**76**:174-187

[59] Liu J, Lin M, Han P, Yao G, Jiang H. Biosynthesis progress of high-energy-density liquid fuels derived from terpenes. Microorganisms. 2024;**12**(4):706

Chapter 3

Phorbol Esters in the *Euphorbiaceae* Family (*Croton*, *Euphorbia*, and *Sapium*)

Arturo Cano-Flores

Abstract

The *Euphorbiaceae* family (229 genera and 6974 species) biosynthesizes particularly diterpenes with a structural variety, where phorbol ethers (tigliane diterpenes) stand out, which present a functionalized tetracyclic ring system, with phorbol isolated from *Croton tiglium* being the most representative compound in this class of secondary metabolites. Different phorbol derivatives have been described, where most of the tigliane derivatives have hydroxyl groups esterified with different saturated and unsaturated long-chain aliphatic fatty acids to naturally produce mono-, di-, and triesters, which have been described in the form of aglycones.

Keywords: *Euphorbiaceae*, *Croton*, *Euphorbia*, *Sapium*, *Jatropha*, phorbol esters

1. Introduction

The present bibliographic review aims to highlight the great chemical diversity of phorbol esters in the *Euphorbiaceae* family, focusing on the genera *Croton*, *Euphorbia*, and *Sapium*. These genera include traditional medicinal plants against many diverse diseases and physical disorders. In addition, they are a rich and varied source of secondary metabolites, showing a wide spectrum of biological activities against diseases of "current society," such as cancer, HIV, rheumatoid fever, SARS-CoV-2, COVID-19, among others. The Euphorbiaceae family is widely distributed in Mexico, where the largest number of species belong to the genera *Euphorbia*, *Croton*, and *Sapium*, with a large endemism of species in these genera. Therefore, in Mexico, it is an urgent task to know and inventory the biological diversity and to assess its importance as a source of chemically and biologically novel secondary metabolites in this family. The information in this work was gathered from several scientific databases, including PubMed, Google Scholar, Scopus, Science Direct, and SciFinder. ChemDraw software was utilized to draw chemical structures.

1.1 Generalities of phorbol esters

Throughout our history, nature has been the main source of chemical compounds useful for preventing, curing, or alleviating several types of diseases. Therefore, it is important to continue researching different living beings (plants, algae, fungi, and marine organisms) to increase the source and availability of biologically active chemical substances. Within vascular plants is the *Euphorbiaceae* family, which is formed in diverse way, both in general and in species, as well as in secondary metabolites (terpenoids, flavonoids, and alkaloids, among others). In addition, due to its traditional uses in the ethnomedicine of many countries and the wide diversity and high toxicity of its secondary metabolites, it has been the subject of study in recent decades [1, 2]. An important characteristic of most *Euphorbiaceae* species is their high toxicity, due to the secretion of a milky latex that is highly irritating to the skin and has carcinogenic effects. The compounds known as phorbol esters (phorboids) are the main cause of the toxic activity of this family, because they "mimic" the action of diacylglycerols (DAG) and arachidonic acid, substances that activate the protein kinase C (PKC) enzyme family, an enzyme that regulates different signal transduction pathways and other cellular metabolic activities, such as the biosynthesis of proteins, DNA, and polyamines, and is also involved in cell differentiation processes and gene expression, among other processes [3–5].

Phorboloids are found naturally in many species of the families *Euphorbiaceae* and *Thymelaeaceae*. The term "phorbol" describe a class of secondary metabolites, which have as their backbone the diterpenes tigliane, ingenane, and daphnane (**Figure 1**) [3, 4]. Phorbol esters are "polycyclic diterpenes hydroxylated at vicinal carbons (C-12 and C-13) esterified with fatty acids with different functionalities" [6–10]. Such compounds have been reported in various plant species of the genera *Croton* (*C. spareiflorus*, *C. tigilium*, *C. ciliatoglandulifer*), *Euphorbia* (*E. cocrulescence*, *E. ticulli*, and other species), *Sapium* (*S. indicum* and *S. japonicum*), and *Jatropha* (*Jatropha curcas*) [3]. The first phorbol ester isolated from *Croton tiglium* plants is known as TPA (4β-12-O-tetradecanoylphorbol-13-acetate). It is the most active phorbol ester and is characteristic of different *Croton* species [6]. In this work, we describe the main esters with a tiglian skeleton present in three genera of the Euphorbiaceae family.

Among all the diterpenoids isolated from the Euphorbiaceae family, there are the skeletons of the "higher diterpenoids," which are cyclized in a classical "accordion-like" manner from the precursor, to generate several cyclic structures.

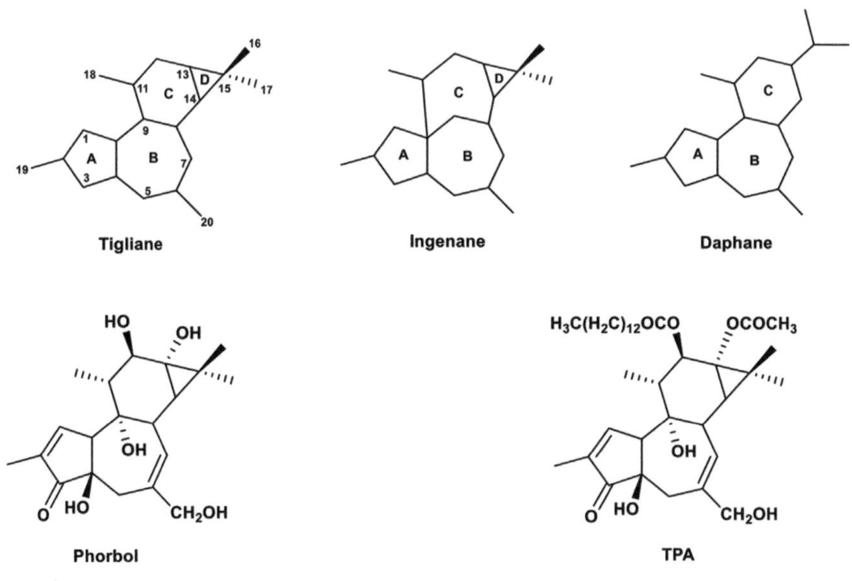

Figure 1.
Basic skeletons of the diterpenes tigliane, ingenane and daphane.

Higher diterpenoids possess three six-membered rings (6/6/6) in their base structure (polycyclic diterpenes). Among all the polycyclic diterpenoids isolated from the Euphorbiaceae family, there are skeletons of labdane, abietane, atisane, kaurane, isopimarane, rosane, and dolabrane (**Figure 2**) [7–10].

Macrocyclic diterpenoids, also known as "lower terpenoids," are compounds that have served as taxonomic markers in the *Euphorbiaceae* and *Thymelaeaceae* families. Head-to-tail cyclization in geranylgeranyl pyrophosphate (PGG) leads to the formation of casbane, cembrane, rharnnofolane, gaditanone, ingenane, ingol, jatrophane, jatropholane, lathyrane, cyclomyrsinol, myrsinol, premyrsinol, paraliane, pepluane, presegtane, segetane, cyclojatrophane, and expoxyjatropholane [7–10].

This paper describes the diterpenes derived from phorbol isolated from three genera (*Croton*, *Euphorbia*, and *Sapium*) and their importance as a source of secondary metabolites with high potential as prototypes for new medicines worldwide. In Mexico, these genera have many plant species, many of which are endemic and have not been biologically described, and lack chemical studies.

1.2 General structural characteristics of tigliane esters

Phorbol esters are lipophilic compounds, found especially in latex seeds of the *Euphorbiaceae* and *Thymelaeaceae* families, where their concentration is much lower than that of the dominant triterpenoids, triglycerides, and fatty acids. They are formed by cyclization between the C-8 and C-14 positions of the lathran skeleton. Phorbol esters are tetracyclic compounds (5/7/6/3), where the fusion of the A/B rings is generally *trans*, while the B/C and C/D rings are *trans* and *cis*, respectively. Most tigliane diterpenoids are hydroxylated at the C-4, C-9, C-13, and C-20. In ring A, there is an α,β-unsaturated carbonyl group with $\Delta^{1(2)}$, and the olefin bond can be $\Delta^{6(7)}$. The carbonyl group in ring A of these compounds is characteristic of this type of secondary metabolite [7, 9–11].

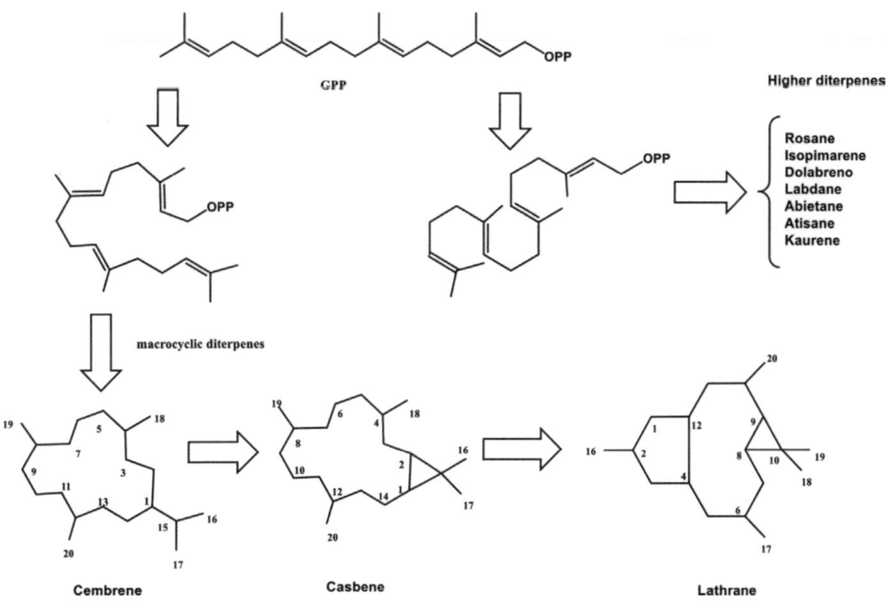

Figure 2.
Formation of the lathrane skeleton, the precursor of phorbol esters.

In the skeletal structure of tigliano, there are three regions of high reactivity: the α,β-unsaturated carbonyl system in ring A, the allylic alcohol in ring B (C-6, C-7, and C-20), and the gem-diol system (C-12 and C-13). The hydroxyl group at C-9 shows neither chemical nor biological reactivity [6–9]. Therefore, it is possible to associate its biological activity with these three centers of chemical reactivity, as is the case with tigilanol tiglate (2), a natural product obtained from *Fontainea pricosperma* (Euphorbiaceae), which is in clinical trials for the treatment of a wide range of cancers (**Figure 3**) [12, 13].

Different phorbol derivatives have been described, including 12-deoxyphorbol, 4-deoxyphorbol, 4-deoxy(4α)phorbol, 4,12-dideoxyphorbol, 4,12-dideoxy(4α) phorbol, 4,20-dideoxyphorbol, 4,20-dideoxy(α)phorbol, and 12,20-dideoxyphorbol. 20-deoxyphorbol, 4,12,20-trideoxyphorbol, and 16 or 17 substituted derivatives. Many derivatives of phorbol esters are hydroxylated at C-12, C-13, and C-20 positions that can be esterified with different carboxylic acids: acetic, isobutyric, tiglic, 2-methylbutyric, benzoic, 2-methylaminobenzoic, or saturated and unsaturated long-chain aliphatic acids. Tiglian derivatives occur naturally as 13-monoesters, diesters (C-12, C-13), (C-13, C-16), (C-12, C-20), and triesters (C-12, C-13, C-20), (C-13, C-16, C-20) [7, 9, 10].

Esterification is a usual form of substitution in diterpenes and triterpenes, which has different effects on skeleton activity [6, 7]. In phorbol esters, the esterification of long-chain saturated and unsaturated fatty acids at the C-12, C-13, and C-20 positions is a common structural feature associated with biological activity. For instance, the esterification of tigliane diterpenoids with long-chain fatty acids at the C-13 position selectively enhances their antiviral activity against HIV and their cytotoxic effects. Overall, it can be concluded that the carbonyl group, hydroxyl (OH) groups, and the 13-acyl type groups in the tigliane structure are key features essential for its

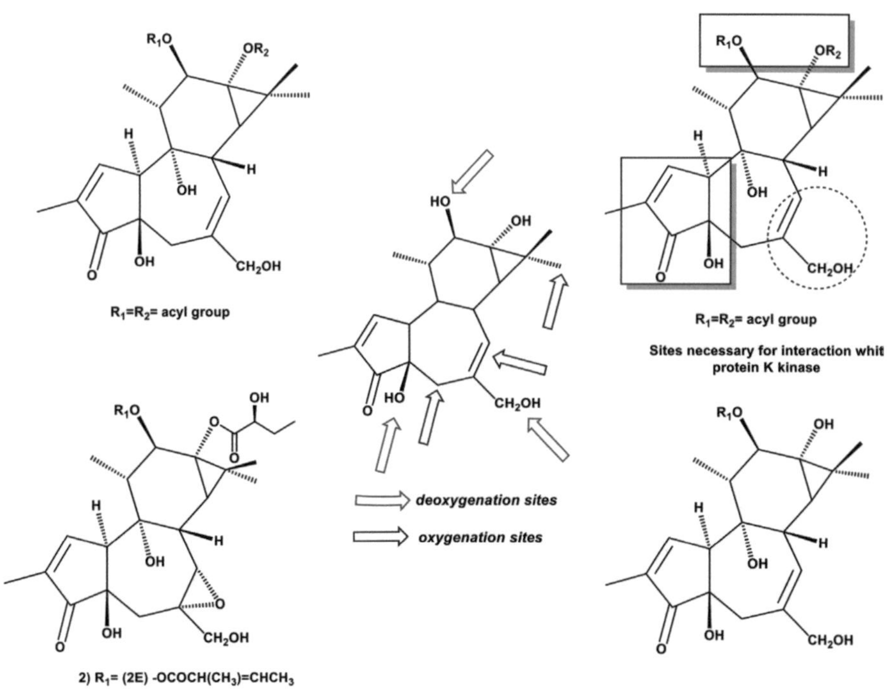

Figure 3.
The basic structure correlates with the biological activity of phorbol ester derivatives.

biological activity [13–15]. One of the most significant structural differences between phorbol ethers of the *Euphorbiaceae* and *Thymelaeaceae* families lies in the high degree of oxygenation in the B ring [11].

So far, few derivatives of phorbol esters in glycosylated form have been described. For example, 13-deoxygenated tigliane-skeletal glycosides and 4-deoxygenated rhamnofolans, which are biogenetically related, have been identified from the root of *E. wallichi* (**Figure 4**) [9, 10, 16].

1.3 Genus *Croton*

The genus name "Croton" (kroton, thick) is derived from the thick, smooth seeds found in most of the subfamily Crotonoideae of the Euphorbiaceae family, which comprises approximately 1300 species of trees, shrubs, and herbs, which are widely distributed in tropical and subtropical regions of the world. Several species within the genus *Croton* are aromatic, indicating they contain volatile oil components. Like many members of the Euphorbiaceae family, *Croton* species may produce latex, which can appear red in certain varieties. This characteristic is often linked to their medicinal properties. Various *Croton* species have been utilized as folk remedies for treating a range of conditions, including digestive tract diseases, abscesses, inflammation, and malaria, particularly in countries across Africa, South Asia, and South America. Secondary metabolites isolated from distinct species of this genus include diterpenoids, sesquiterpenoids, cyclopeptides, limonoids, and tropane alkaloids [17]. In addition to phenolic substances, among which flavonoids, lignoids, and proanthocyanidins predominate, with a wide range of biological activities, such as cytotoxic, anti-inflammatory, antioxidant [17–19], anti-VIH [20–22], antimicrobial, and others [23]. **Table 1** lists the different phorbol esters and derivatives isolated from distinct species of the genus *Croton* (**Figures 5–7**).

1.4 Genus *Euphorbia*

The Euphorbiaceae family, one of the largest groups of higher plants, includes approximately 50 tribes, 300 genera, and about 7500 species, exhibiting the highest

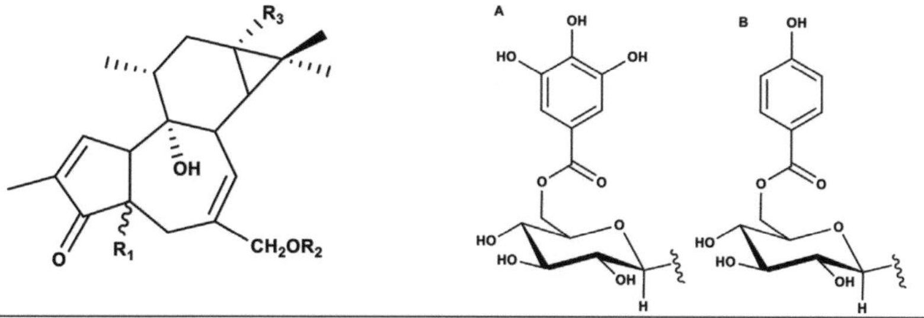

Compound	R_1	R_2	R_3
Euphorwallside A (**3**)	βH	β-D-glucopiranoside	H
Euphorwallside B (**4**)	αH	β-D-glucopiranoside	H
Euphorwallside C (**5**)	βH	β-D-glucopiranoside	H
Euphorwallside C (**6**)	αH	6'-galloly-β-D-glucopiranoside (A)	H
Euphorwallside C (**7**)	αH	6'-galloly-β-D-glucopiranoside (A)	H
Euphorwallside F (**8**)	βOH	β-D-xylopiranoside	OAc
Euphorwallside G (**9**)	βOH	6'-p-hydroxybenzoyl-β-D-glucopiranoside (B)	OAc

Figure 4.
Examples of glycosylated derivatives of phorbol esters.

Specie	Compound (structure)	Reference
C. tiglium	Phorbol (1)	[9, 10]
C. tiglium	Phorbol 12-myristate (**10**)	[24]
C. tiglium	Phorbol 12-acetate (**11**)	[24]
C. tiglium	Phorbol 12,13-diacetate (PDA, **12**)	[24]
C. tiglium	Phorbol 12-acetate 13-tigliate (**13**)	[25]
C. tiglium	Phorbol 12-acetate 13 decanoate (**14**)	[25–27]
C. tiglium	Phorbol 12-acetate 13-dodecanoate (**15**)	[27, 28]
C. tiglium	Phorbol 12-tiglate 13-acetate (**16**)	[26]
C. tiglium	Phorbol 12-decanoate 13-acetate (**17**)	[26]
C. tiglium	Phorbol 12-dodecanoate 13-acetate (**18**)	[26]
C. sparciflorus		[29]
C. tiglium	Phorbol 12-tetradecanoate 13-acetate (**19**)	[25–27]
C. tiglium	Phorbol 12-hexadecanoate 13-acetate (**20**)	[27]
C. tiglium	Phorbol 12,13-dibutyrate (PDB, **21**)	[24]
C. tiglium	Phorbol 12-butyrate 13-dodecanoate (**22**)	[9, 30]
C. tiglium	Phorbol 12-tiglate13-propionate (**23**)	[24]
C. tiglium	Phorbol 12-tiglate 13-isobutyrate (**24**)	[26, 27]
C. tiglium	Phorbol 12-tiglate 13-butyrate (**25**)	[9]
C. tiglium	Phorbol 12-tiglate 13-(2-methyl)-butyrate) (**26**)	[25, 27]
C. tiglium	Phorbol 12-tiglate 13-octenoate (**27**)	[9, 30]
C. tiglium	Phorbol 12-tiglate 13-decanoate (**28**)	[30–32]
C. tiglium	Phorbol 12-tiglate 13-dodecanoate (**29**)	[9, 30]
C. tiglium	Phorbol 12-(2-methyl)-butyrate 13-isobutyrate (**30**)	[27, 28]
C. tiglium	Phorbol 12-(methyl)-butyrate 13-tiglate (**31**)	[27]
C. tiglium	Phorbol 12-(2-methyl)-butyrate 13-octenoate (**32**)	[9, 30]
C. tiglium	Phorbol 12α-(2-methylbutyrate 13-decanoate (**33**)	[28, 30]
C. tiglium	Phorbol 12-(2-methyl)-butyrate 13-dodecanoate (**34**)	[27, 30]
C. tiglium	Phorbol 12,13-dibenzoate (**35**)	[24]
C. tiglium	Phorbol 12-decanoate 13-(2-methyl)-butyrate (**36**)	[25]
C. tiglium	Phorbol 12, 13-didecanoate (**37**)	[24]
C. tiglium	Phorbol 13,20-diacetate (**38**)	[24]
C. tiglium	Phorbol 13-acetate 20-oleate (**39**)	[27]
C. tiglium	Phorbol 13-acetate 20-linoleate (**40**)	[25, 27]
C. tiglium	Phorbol 13-tiglate 20-linoleate (**41**)	[25]
C. tiglium	Phorbol 12,13,20-triacetate (**42**)	[24, 26]
C. tiglium	Phorbol 12-tetradecanoate 13,20-diacetate (**43**)	[26]
C. saparciflorus	Phorbol |12-dodecanoate 13-acetate 20-linolenate (**44**)	[29]
C. tiglium	Phorbol 4α,12,13-didecanoate (**45**)	[24]

Specie	Compound (structure)	Reference
C. tiglium	5,6-didehydro-7-oxophorbol 12-acetate 13-(2-methyl)-butyrate (**46**)	[33]
C. tiglium	5,6-didehydro-7-oxophorbol 12-acetate 13-(2-methyl)-propanoate (**47**)	[33]
C. tiglium	5,6-didehydro-7-hydroxyphorbol 12-acetate 13-(2-methyl)-propanoate (**48**)	[33]
C. tiglium	5,6-didehydro-7-hydroperooxyphorbol 12-decanoate 13-acetate (**49**)	[34]
C. ciliatoglandulifer	Phorbol 12-[(2R)-N,N-dimethyl-3-methylbutanoate] 13-acetate (**50**)	[35]
	4-deoxyphorbol and 4-deoxy-(4α)-phorbol	
C. tiglium	4-deoxy-(4α)-phorbol 12-tiglate 13-acetate (**51**)	[27]
C. tiglium	4-deoxy-(4α)-phorbol 12-tiglate 13-isobutyrate (**52**)	[27]
C. tiglium	4-deoxy-(4α)-phorbol 12-tiglate 13-(2-methyl)-butyrate (**53**)	[27]
C. tiglium	4-deoxy-(4α)-phorbol 12-tiglate 13-phenylacetate (**54**)	[27]
C. tiglium	4-deoxy-(4α)-phorbol 12-tiglate 13-decanoate (**55**)	[27]
C. tiglium	4-deoxy-(4α)-phorbol 13-acetate 20-linoleate (**56**)	[27]
C. tiglium	4-deoxy-(4α)-phorbol 13-acetate 20-oleate (**57**)	[27]
C. ciliatoglandulifer	4-deoxyphorbol-12-(2R)-N,N-dimethyl-3-methylbutanoate 13-acetate (**58**)	[35]
C. ciliatoglandulifer	4-deoxyphorbol-12-(2S)-N,N-dimethyl-3-methylbutanoate 13-acetate (**59**)	[35]
C. ciliatoglandulifer	4α-deoxyphorbol 12-(3-methyl)-2-butenoate 13-acetate (**60**)	[35]
C. alienus	4α-deoxyphorbol-12,20-didecanoate 13-acetate (alienusolina, **61**)	[36]

Table 1.
Phorbol esters and their derivatives are extracted from different species of Croton.

species diversity across various habitats [7]. The genus *Euphorbia* is found worldwide, thriving in both tropical and temperate regions. Its members display a wide range of morphologies, from small annual and perennial herbaceous plants to woody shrubs, lianas, trees, and even large desert succulents [37]. Various parts of *Euphorbia* species, such as whole plants, stems, leaves, latex, roots, and seeds, have been extensively studied for their chemical and pharmacological properties. Several species within this genus hold significant economic value, including *E. tetragona* (which produces lower rubber), *E. antisyphilitica* (source of candelilla wax), and *Euphorbia resinifera* (which produces euphorbium). Chemical studies on various *Euphorbia* species primarily focus on their latex, which is a milky white or pungent liquid released when the plant is cut or damaged. This latex can be extremely irritating to the skin [37–39]. It possesses a range of interesting biological and pharmacological properties, including antibacterial, antioxidant, anti-free radical, cytotoxic, antitumor, anti-inflammatory, healing, hemostatic, antiangiogenic, insecticidal, antiadipogenic, genotoxic, and mutagenic effects [1, 40–42]. However, the role of other latex components, such as phenolic compounds, alkaloids, saponins, and flavonoids, remains unknown, which limits the application of latex [38, 39]. Many of the diterpenes isolated from the genus show different biological activities, such as anticancer [43–46], anti-HIV [40, 47, 48], anti-inflammatory [49, 50], antimicrobial, antifungi [51], antimalarial, K^+ Channel inhibition [52, 53], inhibitory effects on α-glucosidase [54], antiviral [40, 47, 48],

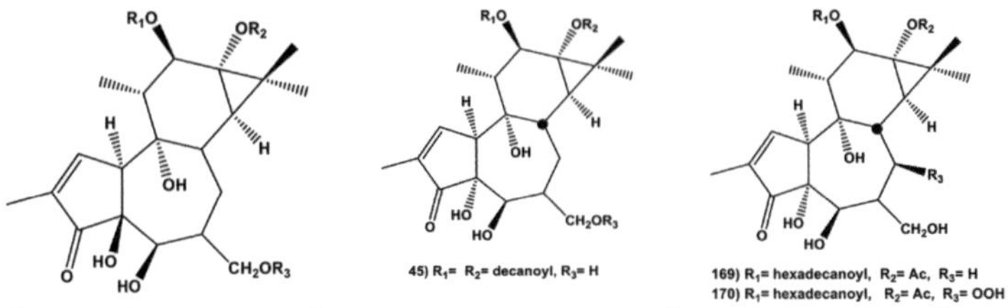

Compound	R_1	R_2	R_3
1	H	H	H
10	myristyl	H	H
11	H	Ac	H
12	Ac	Ac	H
13	Ac	tigloyl	H
14	Ac	decanoyl	H
15	Ac	dodecanoyl	H
16	tigloyl	Ac	H
17	decanoyl	Ac	H
18	dodecanoyl	Ac	H
19	tetradecanoyl	Ac	H
20	hexadecanoyl	Ac	H
21	butyroyl	butyroyl	H
22	butyroyl	dodecanoyl	H
23	tigloyl	propionoyl	H
24	tigloyl	isobutynoyl	H
25	tigloyl	butyroyl	H
26	tigloyl	(2-methyl)-butyroyl	H
27	tigloyl	octenoyl	H
28	tigloyl	decanoyl	H
29	tigloyl	dodecanoyl	H
30	(2-methyl)-butyroyl	isobutyroyl	H
31	(2-methyl)-butyroyl	tigloyl	H
32	(2-methyl)-butyroyl	octenoyl	H
33	(2-methyl)-butyroyl	decanoyl	H
34	(2-methyl)-butyroyl	dodecanoyl	H
35	benzoyl	benzoyl	H
36	decanoyl	(2-methyl)-butyroyl	H
37	decanoyl	decanoyl	H
38	H	Ac	Ac
39	H	Ac	oleoyl
40	H	Ac	linoleoyl
41	H	tigloyl	linoleoyl
42	Ac	Ac	Ac
43	tetradecanoyl	Ac	Ac
44	dodecanoyl	Ac	linoleoyl
45	decanoyl	decanoyl	H
50	(2R)-N,N-dimethyl-3-methylbutanoyl	Ac	H
63	Ac	2,4,6,8,10-tetradecapentaenoyl	H
64	(2Z,4E)-2.4-octadienoyl	Ac	H
65	(2Z,4E)-2,4,6-decatrienoyl	Ac	H
66	2,4,6,8-tetradecatetraenoyl	Ac	H
67	2,4,6,8-tetradecatetraenoyl	Ac	H
68	(2-methyl)-butyroyl	angeloyl	H
69	H	Ac	glucosyl
70	isobutyroyl	Ac	angeloyl
71	(2Z,4E)-2,4-octadienoyl	Ac	Ac
165	N-methylaminobenzoyl	Ac	H
166	2,4,6-decatrienoyl	Ac	H
167	octadecanoyl	Ac	H
168	octadecenoyl	Ac	H
169	hexadecanoyl	Ac	H
170	hexadecanoyl	Ac	H
171	hexadecanoyl	Ac	H

Figure 5.
*Chemical structures of phorbol esters (***1, 10–45, 63–71** *and* **165–171***).*

antifeedant [55], neuroprotection effects and multiple drug resistance (MDR) reversal [56–59], antiviral activity against SARS-CoV-2 [60]. **Table 2** describes the phorbol esters and their different derivatives isolated from different genera of *Euphorbia* species (**Figures** 5, 7–12].

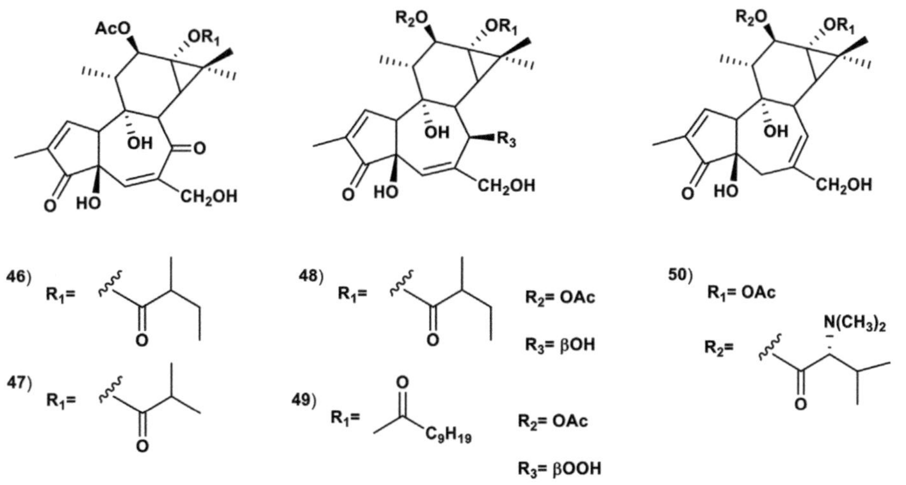

Figure 6.
*Chemical Structure of Phorbol Esters and Their Derivatives (**46**–**50**).*

1.5 Genus *Sapium*

The genus *Sapium* is known for its diverse biological activities and is utilized in the traditional medicine of various indigenous groups around the world. This genus is distributed across regions, including Malaysia, Mexico, Central America, the Caribbean, tropical South America, Southern Brazil, and temperate South America. *Sapium* species, for example, *S. sebiferum*, can be used to extract industrial oils, manufacture paper, cardboard, and furniture materials. In addition, *S. sebiferum* is used as an ornamental plant due to its long life and colorful foliage. *S. glandulosum*, *S. indicum*, and *S. sebiferum* have toxic effects and are used as poisons for birds and fish [96, 97].

The chemical analysis of various species belonging to the genus *Sapium* has identified several secondary metabolites, including terpenoids, phenylpropanoids, flavonoids, tannins, steroids, and alkaloids. Among these, terpenoids, phenylpropanoids, and tannins are the primary constituents [96, 97]. Pharmacological studies, both *in vitro* and *in vivo*, have demonstrated that the extracts and pure compounds possess significant antihypertensive [97], cytotoxic [98], anti-inflammatory [99], antioxidant [99] antibacterial [98, 100], antidiabetic, and molluscicidal [96, 97]. **Table 3** provides a list of phorbol esters and their derivatives that have been isolated from the genus *Sapium* (**Figures** 5, 7, **8**, and **13–15**).

1.6 Plant species of the *Euphorbiaceae* family in Mexico

In Mexico, the *Euphorbiaceae* are comprised of 43 genera and 782 species. The largest genera are *Euphorbia* with 241 species (31%), *Croton* (124, 6%), *Acalypha* (108, 14%), *Jatropha* (48, 6%), and *Phyllanthus* (41 spp., 5%). Together, they represent 72% of the Euphorbiaceae species known from Mexico. Most species are endemic to the country, where they represent a center of diversity for many genera. Despite its significant importance both systematically and floristically, knowledge of the *Euphorbiaceae* family in Mexico is still extremely poor, considering that so far less than 25% of the species have been subjected to a revision [113].

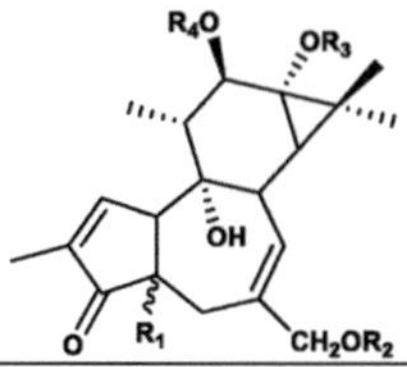

Compound	R_1	R_2	R_3	R_4
51	αH	H	Ac	tigloyl
52	αH	H	isobutyroyl	tigloyl
53	αH	H	(2-methyl)-butyroyl	tigloyl
54	αH	H	phenylacetoyl	tigloyl
55	αH	H	decanoyl	tigloyl
56	αH	linoleoyl	Ac	H
57	αH	oleoyl	Ac	H
58	αH	H	Ac	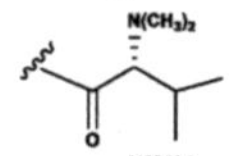
59	αH	H	Ac	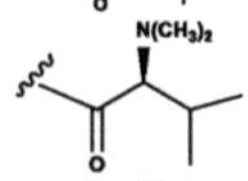
60	αH	H	Ac	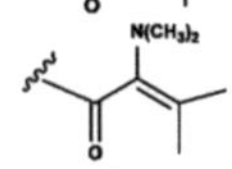
114	βH	H	2,4-octadienoyl	Ac
115	βH	H	1-octenoyl	Ac
116	βH	H	2,4-octadienoyl	Ac
117	βH	H	(2Z, 4E)-2,4,6-decatrienoyl	Ac
118	βH	H	2,4,6-dodecatrienoyl	Ac
119	βH	H	2,4,6,8-dodecatetraenoyl	Ac
120	βH	H	2,4,6,8-tetradecatetraenoyl	Ac
121	βH	H	2,4,6,8,10-tetradecapentaenoyl	Ac
122	βH	H	isobutyroyl	Isobutyroyl
123	βH	H	isobutyroyl	2,4-decadienoyl
124	βH	H	isobutyroyl	2,4-decadienoyl
125	βH	Ac	Ac	(2Z, 4E)-2,4-octadienoyl
126	βH	Ac	isobutyroyl	benzoyl
127	βH	Ac	isobutyroyl	isobutyroyl
128	βH	benzoyl	Ac	isobutyroyl
129	αH	H	Ac	(2Z, 4E)-2,4,6-decatrienoyl
130	αH	H	isobutyroyl	Isobutyroyl
131	αH	H	isobutyroyl	(2,3-dimethyl)-butyroyl
132	αH	H	2,4,6,8-tetradecatetraenoyl	Ac
133	αH	Ac	isobutyroyl	isobutyroyl
134	αH	Ac	isobutyroyl	benzoyl
177	βH	H	H	H
178	βH	H	Ac	N-methylantranoyl
179	βH	H	Ac	2,4,6-decatrienoyl
180	βH	H	(2-methyl)-butanoyl	angeloyl
181	αH	H	(2-methyl)-butanoyl	angeloyl
182	βH	H	Ac	Ac
183	αH	H	Ac	N-methylantranoyl
184	αH	H	Ac	hexanoyl

Figure 7.
*Chemical structures of 4-deoxyphorboles and 4-deoxy-(4a)-phorboles (***51–60, 114–134** *and* **177–184***).*

Specie	Compound (structure)	Reference
E. fischeriana	Phorbol 13-acetate (**61**)	[61, 62]
E. fischeriana	Phorbol 4β-hydroxy (**62**)	[62]
E. tirucalli	Phorbol 12-acetate 13-(2,4,6,8,10)-tetradecapentaenoate (**63**)	[63, 64]
E. tirucalli	Phorbol 12-[(2Z, 4E)-2,4-octadienoate] 13-acetate (**64**)	[64]
E. tirucalli	Phorbol 12-[(2Z, 4E)-2,4,6-decatrienoate] 13-acetate (**65**)	[64]
E. tirucalli	Phorbol 12-(2,4,6,8-tetradecatetraenoate) 13-acetate (**66**)	[64]
E. tirucalli	Phorbol 12-(2,4,6,8,10-tetradecapentaenoate) 13-acetate (**67**)	[65]
E, tirucalli	Phorbol 12-(2-methylbutyrate) 13-angelate (**68**)	[64]
E. fischeriana	Fischeroside C (**69**)	[66]
E. franckiana	Phorbol 12-isobutyrate 13-acetate 20-angelate (**70**)	[64]
E. broteri	Phorbol 12-(2Z, 4E)-2,4-oactadienoate 13.20-diacetate (**71**)	[67]
	12-deoxyphorbol	
E. triangularis	12-deoxyphorbol (**72**)	[9]
E. resinifera		[68]
E. fischeriana	12-deoxyphorbol 13-acetate (prostratin, **73**)	[69]
E. triangularis		[70]
E. cornigera		[71]
E. ledienii	12-deoxyphorbol 13-isobutyrate (**74**)	[70]
E. resinífera		[9, 72]
E. triangularis		[9, 70]
E. poisonii		[9]
E. poisonii	12-deoxyphorbol 13-(2-methyl)-butyrate (**75**)	[73]
E. ledienni		[70]
E. coerulescens		[9, 70]
E. fortissima		[9]
E. triangularis		[9]
E. unispina		[9]
E. poisonni	12-deoxyphorbol 13-phenylacetate (**76**)	[73]
E. resinifera		[72]
E. fortissima	12-deoxyphorbol 13-tigliate (**77**)	[9]
E. ledienni		[70]
E. triangularis		[9, 70]
E. helioscopia		[9]
E. poisonni	12-deoxyphorbol 20-hydroxy 13-angelate (**78**)	[9, 74]
E. resinífera		[72]

Specie	Compound (structure)	Reference
E. unisipina		[9]
E. resinifera	12-deoxyphorbol 13-(4-methoxyphenyl)-acetate (**79**)	[72]
E. coerulescens	12-deoxyphorbol 13-heptanoato (**80**)	[75]
E. poisonni	12-deoxyphorbol 20-hydroxy 13(*cis*-9,10-methylene)-undecanoate (**81**)	[74]
E, fortissima	12-deoxyphorbol 13-dodecenoate (**82**)	[75]
E. coerulescens	12-deoxyphorbol 13-dodecanoate (**83**)	[75]
E. fortissima		[9, 10]
E. fischeriana	12-deoxyphorbol 13-hexadecanoate (**84**)	[69, 76]
E. fischeriana	12-deoxyphorbol 13,20-diacetate (**85**)	[10, 66]
E. poisonni	12-deoxyphorbol 13-isobutyrate 20-acetate (**86**)	[9, 10]
E. ledienii		[70]
E. fortissima		[10, 75]
E. triangularis		[70]
E. resinifera		[72]
E. poisonni	12-deoxyphorbol 13-angelate 20-acetate (**87**)	[9, 74]
E.unispina		[9, 10]
E. resinifera		[10, 72]
E. poisonni	12-deoxyphorbol 13-(2-methyl)-butyrate 20-acetate (**88**)	[9, 10]
E. ledienii		[70]
E. fortissima		[9]
E. triangularis		[9, 10]
E. unisipina		[9, 10]
E. coerulescens		[70, 75]
E. fortissima	12-deoxyphorbol 13-tiglate 20-acetate (**89**)	[75]
E. helioscopia		[75]
E. coerulescens		[77]
E. ledienii		[70]
E. triangularis		[10, 70]
E. poisonni	12-deoxyphorbol 13-(phenyl)-acetate 20-acetate (**90**)	[73, 74]
E. resinifera		[72]
E. poisonni	12-deoxyphorbol 13-(p-hydroxyphenyl)-acetate 20-acetate (**91**)	[73, 75]
E. resinifera	12-deoxyphorbol 13-(4-methoxyphenyl)-acetate 20-acetate (**92**)	[72]
E. poisonni	12-deoxyphorbol 13-(acetoxyphenyl)-acetate 20-acetate (**93**)	[75, 78]
E. polyacantha	12-deoxyphorbol 13-octenoate 20-acetate (**94**)	[75]
E. polyacantha	12-deoxyphorbol 13-decadienoate 20-acetate (**95**)	[75]
E. coerulescens	12-deoxyphorbol 13-dodecanoate 20-acetate (**96**)	[75, 79]
E. fortissima	12-deoxyphorbol 13-dodecenoate 20-acetate (**97**)	[9, 75]
E. helioscopia	12-deoxyphorbol 13-dodecadienoate 20-acetate (**98**)	[75, 77]

Specie	Compound (structure)	Reference
E. fischeriana	12-deoxyphorbol 13-(9Z)-octadecenoate 20-acetate (**99**)	[76]
E. cornigera	12-deoxyphorbol 13-acetate 20-benzoate (**100**)	[71]
E. cornigera	12-deoxyphorbol 13-acetate 20-p-methoxybenzoate (**101**)	[71]
E. fischeriana	Ficheroside A (**102**)	[66]
E. fischeriana	Ficheroside B (**103**)	[66]
E. cornigera	12-deoxyphorbol 13-acetate 20-decanoate (**104**)	[71]
E. cornigera	12-deoxyphorbol 13-butanoate 20-decanoate (**105**)	[71]
E. cornigera	12-deoxyphorbol 13-hexanoate 20-decanoate (**106**)	[71]
E. cornigera	12-deoxyphorbol 13-octanoate 20-decanoate (**107**)	[71]
E. cornigera	12-deoxyphorbol 13-decanoate 20-angelate (**108**)	[71]
E. cornigera	12-deoxyphorbol 13-decanoate 20-tiglate (**109**)	[71]
E. cornigera	12-deoxyphorbol 13,20-didecanoate (**110**)	[71]
E. cornigera	12-deoxyphorbol 13-dodecanoate 20-decanoate (**111**)	[71]
E. fischeriana	12-deoxyphorbolaldehyde 13-acetate (**112**)	[76]
E. fischeriana	12-deoxyphorbolaldehyde 13-hexadecanoate (**113**)	[76]
	4-deoxyphorbol and 4-deoxy-(4α)-phorbol	
E. tirucalli	4-deoxyphorbol 12-acetate 13-(2,4-octadienoate) (**114**)	[80, 81]
E. tirucalli	4-deoxyphorbol 12-acetate 13-octenoate (**115**)	[80, 81]
E. tirucalli	4-deoxyphorbol 12-acetate 13-(2,4-decadienoate) (**116**)	[80, 81]
E. tirucalli	4-deoxyphorbol 12-acetate 13-[(2Z,4E)-2,4.6-decatrienoate] (**117**)	[80, 81]
E. tirucalli	4-deoxyphorbol 12-acetate 13-dodecatrienoate (**118**)	[80, 81]
E. tirucalli	4-deoxyphorbol 12-acetate 13-(2,4,6,8-dodecatenoate) (**119**)	[80, 81]
E. tirucalli	4-deoxyphorbol 12-acetate 13-(2,4,6,8-tetradecatretraenoate) (**120**)	[82]
E. tirucalli	4-deoxyphorbol 12-acetate 13-(2,4,6,8,10-tetradecapentenoate) (**121**)	[82]
E. obtusifolia	4-deoxyphorbol 12,13-diisobutyroate (**122**)	[83, 84]
E. prolifera	4-deoxyphorbol 12-(2,4-decadienoate) 13-isobutyrate (**123**)	[85]
E. prolifera	4-deoxyphorbol 12-(2,4,6-decatrienoate) 13-isobutyrate (**124**)	[85]
E. broteri	4-deoxyphorbol 12-(2Z,4E)-octadienoate 13,20-diacetate (**125**)	[56]
E. blingandulosa		[9]
E. rubica	4-deoxyphorbol 12-benzoate 13-isobutyrate (**126**)	[86]
E. obtusifolia	4-deoxyphorbol 12,13-diisobutyrate 20-acetate (**127**)	[83, 84]
E. rubica	4-deoxyphorbol 12-isobutyrate 13-acetate 20-benzoate (**128**)	[86]
E. tirucalli	4-deoxy-(4α)-phorbol 12-[(2Z, 4E)-2,4,6-decatrienoate 13-acetate (**129**)	[9, 10]
E. obtusifolia	4-deoxyphorbol –4-epi-12,13-diisobutyrate (**130**)	[83, 84]
E. aellenii	4-deoxy-(4α)-phorbol 12-[(2,3-dimethyl)-butyrate 13-isobutyrate (**131**)	[9, 10]
E. tirucalli	4-deoxy-(4α)-phorbol 12-acetate 13-(2,4,6,8-tetradecatetraenoate (**132**)	[9, 10]
E. rubica	4-deoxy-(4α)-phorbol 12,13-diisobutyrate 20-acetate (**133**)	[86]
E. rubica	4-deoxy-(4α)-phorbol 12-benzoate 13-isobutyrate 20-acetate (**134**)	[86]

Specie	Compound (structure)	Reference
	4,12-dideoxyphorbol and 4,12-dideoxy-(4α)-phorbol	
E. pithyusa	4,12-dideoxyphorbol 13-(2,3-dimethyl)-butyrate (**135**)	[87]
E. pithyusa	4,12-dideoxyphorbol 13-(2,3-dimethyl)-butyrate 20-acetate (**136**)	[87]
E. pannonica	4,12-dideoxyphorbol 13-isobutyrate 20-benzoate (**137**)	[88]
E. pannonica	4,12-dideoxyphorbol 13-isovalerianate 20-benzoate (**138**)	[88]
E. guyoniana	4,12-dideoxy-(4α)-phorbol 13-hexadecanoate (**139**)	[89]
E. tuckeyana	4,20-dideoxyphorbol-5β-hydroxy-12,13-diisobutyrate (**140**)	[90]
E. tuckeyana	4,20-dideoxyphorbol-5β-hydroxy-12-benzoate 13-isobutyrate (**141**)	[90]
E. nubica	4,20-dideoxyphorbol-5ξ-acetate-12-benzoate 13-isobutyrate (**142**)	[86]
E. prolifera	4,20-dideoxyphorbol-5ξ-hydroxy-12,13-diisobutyrate (**143**)	[85]
E. prolifera	4,20-dideoxyphorbol-5ξ-hydroxy-12-benzoate 13-isobutyrate (**144**)	[85]
	12,20-dideoxyphorbol	
E. unispina	12,20-dideoxyphorbol 13-isobutyrate (**145**)	[72, 91, 92]
E. resinifera		[92, 93]
E. unispina	12,20-dideoxyphorbol 13-angelate (**146**)	[92]
E. resinifera		[72, 92]
	4,12,20-trideoxyphorbol	
E. pityusa	4,12-trideoxyphorbol 13-(2,3-dimethyl)-butyrate (**147**)	[87]
	Other derivates of tigliane esters	
E. grandicornis	12-deoxyphorbol-5-en-7-oxo 13-isobutyrate 16-angelate (**148**)	[94]
E. grandicornis	12-deoxyphorbol-5-en-7β-hydroxy-13-isobutyrate 16-angelate (**149**)	[94]
E. poisonni	12-deoxyphorbol-16-hydroxy (**150**)	[9, 10]
E. poisonni	12-deoxyphorbol-16-hydroxy 13-phenylacetate (**151**)	[95]
E. ledienii	12-deoxyphorbol 13-angelate 16-isobutyrate (**152**)	[70]
E. cooperi		[9, 10]
E. triangularis		[70]
E. cooperi	12-deoxyphorbol 13-tiglate 16-isobutyrate (**153**)	[96]
E. coerulescens	12-deoxyphorbol 13-angelate 16-(2-methyl)-butyrate (**154**)	[70]
E. poisonni	12-deoxyphorbol 13-phenylacetate 16-(2-methyl)-butyrate (**155**)	[95]
E. resinifera	12-deoxyphorbol13-phenylacetate 16-benzoate (**156**)	[72]
E. ledienii	12-deoxyphorbol 13-angelate 16-isobutyrate 20-acetate (**157**)	[70]
E. cooperi		[9, 10]
E. triangularis		[70]
E. cooperi	12-deoxyphorbol 13-tigliate 16-isobutyrate 20-acetate (**158**)	[96]
E. lledienni	12-deoxyphorbol 13-angelate 16-(2-methyl)-butyrate 20-acetate (**159**)	[70, 97]
E. coerulescens		
E. poisonni	12-deoxyphorbol 13-phenylacetate 16-(2-methyl)-butyrate 20-acetate (**160**)	[75]
E. resinifera	12-deoxyphorbol 13-phenylacetate 16-benzoate 20-acetate (**161**)	[72]

Specie	Compound (structure)	Reference
E. poisonni	12-deoxyphorbol 13-phenylacetate 16-(2-methyl)-butyrate 20-(methyl)-butyrate (**162**)	[95]
E. obtusifolia	4,20-dideoxyphorbol 12,13-diisobutyrate 17-aceate (**163**)	[83, 85]
E. obtusifolia	4,20-dideoxyphorbol 12,13-diisobutyrate 17-acetate (**164**)	[83, 85]

Table 2.
Phorbol esters and their derivatives are isolated from different species of the genus Euphorbia.

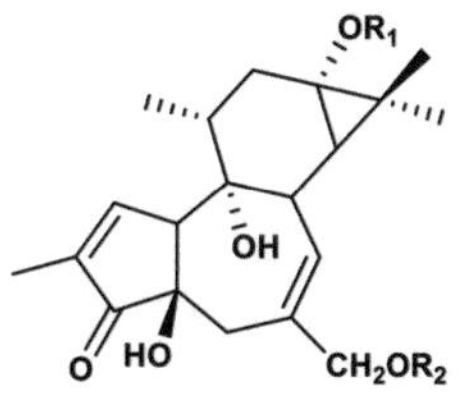

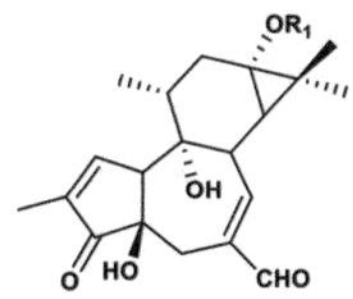

112) R_1= Ac
113) R_1= hexadecanoyl

Compound	R_1	R_2
72	H	H
73	Ac	H
74	isobutyroyl	H
75	(2-methyl)-butyroyl	H
76	phenylacetoyl	H
77	tigloyl	H
78	angeloyl	H
79	(4-methoxyphenyl)-acetoyl	H
80	heptanoyl	H
81	(cis-9,10-methylene)-undecanoyl	H
82	dodecenoyl	H
83	dodecanoyl	H
84	hexadecanoyl	H
85	Ac	Ac
86	isobutyroyl	Ac
87	angeloyl	Ac
88	(2-methyl)-butyroyl	Ac
89	tigloyl	Ac
90	phenylacetoyl	Ac
91	(p-hydroxyphenyl)-acetoyl	Ac
92	(4-methoxyphenyl)-acetoyl	Ac
93	(p-acetoxyphenyl)-acetoyl	Ac
94	octenoyl	Ac
95	decadienoyl	Ac
96	dodecanoyl	Ac
97	dodecenoyl	Ac
98	dodecadienoyl	Ac
99	(9Z)-octadecanoyl	Ac
100	Ac	benzoyl
101	Ac	p-methoxybenzoyl
102	Ac	glucosyl
103	Ac	glucosyl-galloyl
104	Ac	decanoyl
105	butanoyl	decanoyl
106	hexanoyl	decanoyl
107	octanoyl	decanoyl
108	decanoyl	angeloyl
109	decanoyl	tigloyl
110	decanoyl	decanoyl
111	dodecanoyl	decanoyl
172	benzoyl	H
173	tetradecanoyl	H
174	3-tetradecenoyl	H
175	3,5-tetradecadienoyl	H
176	tetradecatrienoyl	H

Figure 8.
*Chemical structures of 12-deoxyphorbol (***72–111, 112–113, 172–176***).*

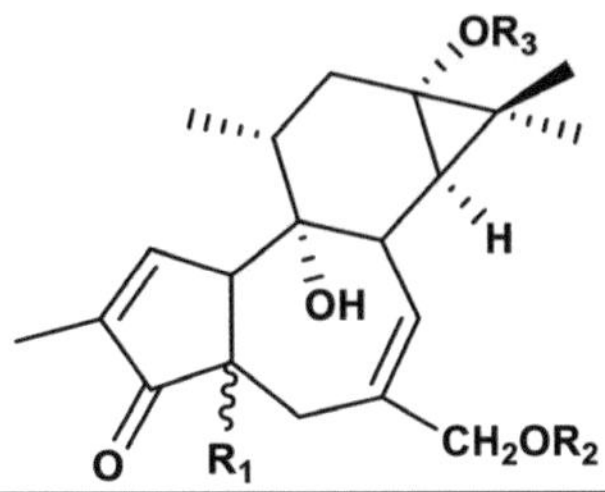

Compound	R1	R2	R3
135	βH	H	(2,3-dimethyl)-butyroyl
136	βH	Ac	(2,3-dimethyl)-butyroyl
137	βH	benzoyl	isobutyroyl
138	βH	benzoyl	isovalerianoyl
139	αH	H	hexadecanoyl

Figure 9.
*Chemical structures of 4,12-dideoxyphorbol and 4,12-dideoxy-(4α)-phorbol derivatives (***135–139***).*

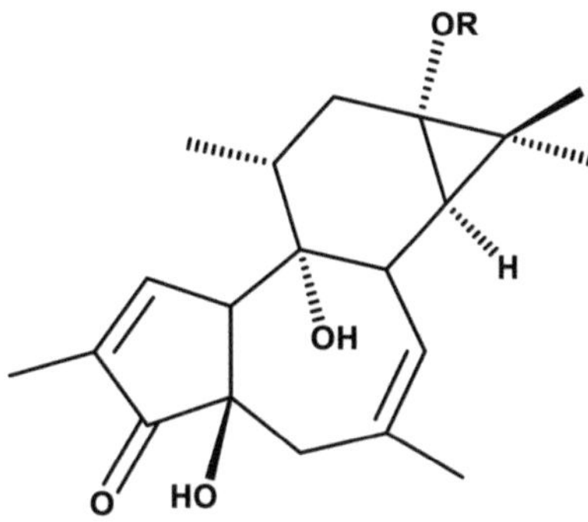

145) R= isobutyroyl
146) R= angeloyl

Figure 10.
*Chemical structures of 12,20-dideoxyphorbol (***145–146***).*

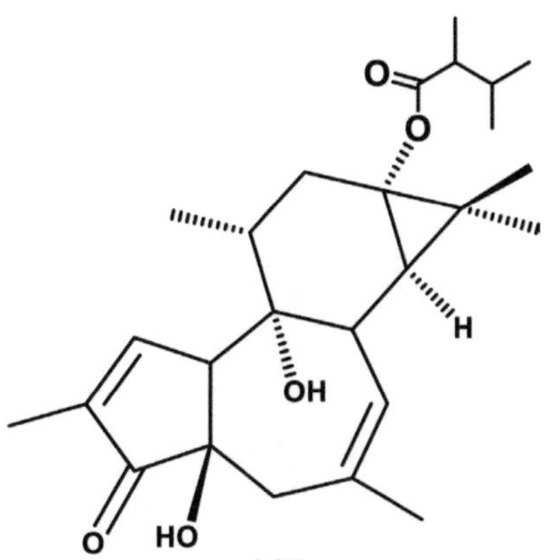

147

Figure 11.
*Chemical structure of 4,12,20-trideoxyphorbol (***147***).*

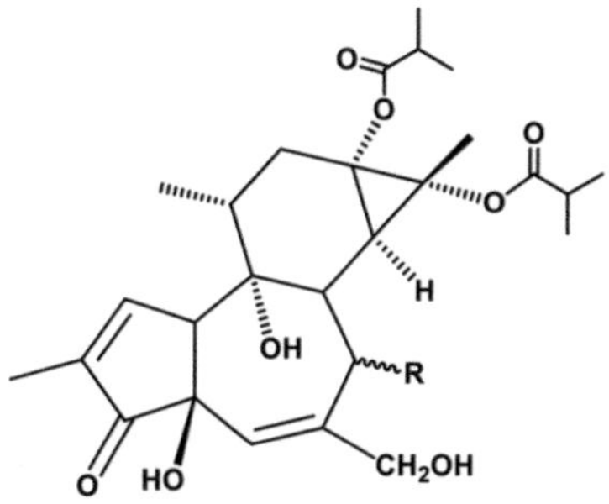

148) R= O
149) R= βOH

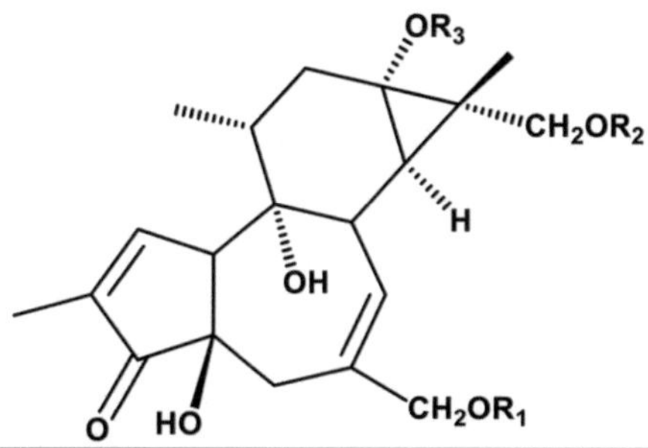

Compound	R_1	R_2	R_3
150	H	H	H
151	H	H	phenylacetyl
152	H	isobutyroyl	angeloyl
153	H	isobutyroyl	tigloyl
154	H	(2-methyl)-butyroyl	angeloyl
155	H	(2-methyl)-butyroyl	phenylacetyl
156	H	benzoyl	phenylacetyl
157	Ac	isobutyroyl	angeloyl
158	Ac	isobutyroyl	tigloyl
159	Ac	(2-methyl)-butyroyl	angeloyl
160	Ac	(2-methyl)-butyroyl	phenylacetyl
161	Ac	benzoyl	phenylacetyl
162	(2-methyl)-butyroyl	(2-methyl)-butyroyl	phenylacetyl

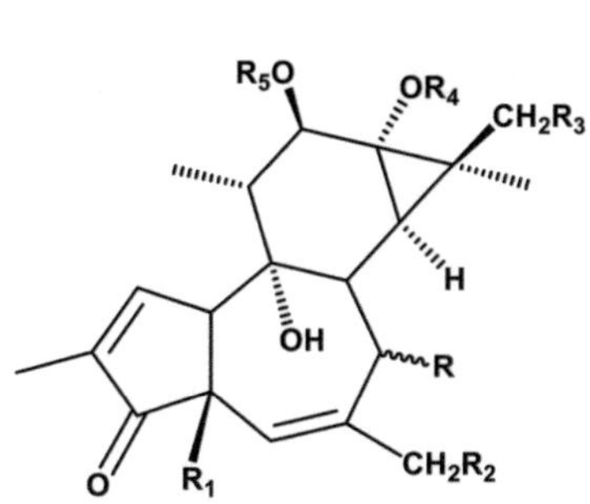

163) R_1= βH, R_2= H, R_3= OAc, R_4= R_5= isobutyroyl
164) R_1= βH, R_2= OH, R_3= OAc, R_4= R_5= isobutyroyl

Figure 12.
*Chemical structures of phorbol esters with 16 or 17 substitutions (***148–149, 150–162, 163–164***).*

Specie	Compound (structure)	Reference
S. indicum	Phorbol 12-(N-methylaminobenzoate) 13 acetate (sapintoxin D, **165**)	[101, 102]
S. japonicum	Phorbol 12-(2,4,6-decatrienoate) 13-acetate (**166**)	[9]
S. sebiferum	Phorbol 12-hexadecanoate 13-acetate (**167**)	[103, 104]
S. sebiferum	Phorbol 12-octadecenoate 13-acetate (**168**)	[104]
S. sebiferum	Phorbol 7-oxo-5-ene-12-hexadecanoate 13-acetate (**169**)	[105]
S. sebiferum	Phorbol 7β-hydroperoxide-5-en-12-hexadecanoate 13-acetate (**170**)	[105]
S. sebiferum	Phorbol 6α,7α-epoxy 12-hexadecanoate 13-acetate (**171**)	[104]
	12-deoxyphorbol	
S. sebiferum	12-deoxyphorbol 13-benzoate (**172**)	[106]
S. sebiferum	12-deoxyphorbol 13-tetradecanoate (**173**)	[104]
S-sebiferum	12-deoxyphorbol 13-(3-tetradecenoate) (**174**)	[104]
S. sebiferum	12-deoxyphorbol 13-(3,5-tetradecadienoate) (**175**)	[104]
S. sebiferum	12-deoxyphorbol 13-tetradecatrienoate (**176**)	[104]
	4-deoxyphorbol and 4-deoxy-(4α)-phorbol	
S. hippomane	4-deoxyphorbol (**177**)	[107]
S. indicum	4-deoxyphorbol 12-(N-methylantraniloate) 13-acetate (sapintoxin A, **178**)	[108–110]
S.sebiferum		
S. indicum	4-deoxyphorbol 12-(2,4,6-decatrienoate) 13-acetate (sapatoxin A, **179**)	[106, 111]
S. hippomane	4-deoxyphorbol 12,13-di(2′-methyl-2′-butenoate) (**180**)	[107]
S. hippomane	4-deoxyphorbol 4-epi-12,13-di(2′-methyl-2″-butanoate) (**181**)	[107]
S. hippomane	4-deoxyphorbol 12,13,20-triacetate (**182**)	[107]
S. indicum	4-deoxy-(4α)-phorbol 12-(2′-N-methylanthaniloate)-13-acetate (α-sapinine, **183**)	[108]
Sapium insigne	4-deoxy-(4α)-phorbol 12-hexanoate 13-acetate (**184**)	[112]
S. indicum	4-deoxy-(4α)-phorbol 5,20-dihidroxy 12-(2′-N-methylaminobenzoate] 13-acetate (**185**)	[108]
S. indicum	4-deoxyphorbol 5,20-dihydroxy 12-(2′-N-methylaminobenzoate 13-acetate (sapintoxin B, **186**)	[99, 113]
S. indicum	4-deoxy-5hydroxy 12-(2,4,6-decatrienato) 13-acetate (sapatoxin B, **187**)	[99, 113]
S. indicum	4-deoxy-(4α)-phorbolaldehyde 12-(2′-N-methylaminobenzoate) 13-acetate (**188**)	[99, 113]
S. indicum	4-deoxyphorboladehyde 12-(2′-N-methylaminobenzoate) 13-acetate (**189**)	[108]
S. indicum	4,5,20-trideoxyphorbol 12-(2′-N-methylaminobenzoate) 13-acetate (**190**)	[108]
S. indicum	4α,20-dideoxyphorbol 12-(2′-N-methylamino benzoate) 13-acetate (**191**)	[108]
S. indicum	4-deoxy-(4β)-phorbolaldehyde 12-(2′-N-methylamino benzoate) 13-acetate (**192**)	[109]
S. indicum	4-deoxy-(4α)-phorbolaldehyde 12-(2′-N-methylaminobenzoate) 13-acetate (**193**)	[109]
S. indicum	4,20-dideoxy-(4α)-5-hydroxyphorbol 12-(2′-N-methylaminobenzoate) 13-acetate (**194**)	[99, 109]

Specie	Compound (structure)	Reference
	4,20-dideoxyphorbol and 4,20-dideoxy-(4α)-phorbol	
S. indicum	4,20-dideoxyphorbol 12-(2′-N-methylaminobenzoate) 13-acetate (**195**)	[114]
S. indicum	4,20-dideoxy-(4α)-phorbol 12-(2′-N-methylaminobenzoate) 13-acetate (**196**)	[108]
S. indicum	4,20-dideoxy-(4α)phorbol-5β-hydroxy 12(2′-N-methylaminobenzoate) 13-acetate (**197**)	[108]
S. indicum	4,20-dideoxyphorbol-5β-hydroxy 12-(2′-N-methylaminobenzoate) 13-acetate (sapintoxin C, **198**)	[108, 113]
S. indicum	4,20-dideoxyphorbol-5β-hydroxy 12-(2,4,6-decatrienoate) 13-acetate (sapatoxin C, **199**)	[99, 113]
S. indicum	4,20-dideoxyphorbol-5β-hydroxy 12-tetradecanoate 13-acetate (**200**)	[99, 113]
	derivatives of phorbol esters substituted at positions 16 or 17.	
S. insigne	4-deoxyphorbol 16-acetate (**201**)	[99, 113]
S. insigne	4-deoxyphorbol 12-(2E,4E)-2,4-decadienoate 13-acetate 16-hydroxy (**202**)	[112]
S. insigne	Phorbol 16-hydroxy (**203**)	[99, 113]
S. insigne	Phorbol 16-acetate (**204**)	[112]
S. insigne	Sapinsignoid A (**205**)	[113]
S. insigne	Sapinsignoid B (**206**)	[113]
S. insigne	Sapinsignoid C (**207**)	[113]
S. insigne	Sapinsignoid D (**208**)	[113]
S. insigne	4-deoxy-(4α)-phorbol 12-hexanoate 13-acetate 16-hydroxy (**209**)	[115]
S. insigne	4-deoxy-(4α)-phorbol 12-dodecanoate 13-acetate 16-hydroxy (**210**)	[115]

Table 3.
Phorbol esters and their derivatives are isolated from different species of the genus Sapium.

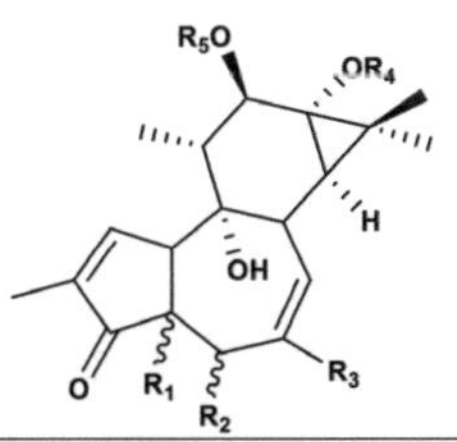

Compound	R_1	R_2	R_3	R_4	R_5
185	αH	βOH	CH_2OH	Ac	N-methylaminobenzoyl
186	βH	βOH	CH_2OH	Ac	N-methylaminobenzoyl
187	βH	βOH	CH_2OH	Ac	decatrienoyl
188	αH	H	CHO	Ac	N-methylaminobenzoyl
189	βH	H	CHO	Ac	N-methylaminobenzoyl
190	H	H	CH_3	Ac	N-methylaminobenzoyl
191	αH	H	CH_3	Ac	N-methylaminobenzoyl
192	βH	H	CHO	Ac	N-methylaminobenzoyl
193	αH	H	CHO	Ac	N-methylaminobenzoyl
194	αH	OH	CHO	Ac	N-methylaminobenzoyl

Figure 13.
*Chemical structures of 4-deoxyphorbol and 4-deoxy-(4α)-phorbol derivatives (***185–194***).*

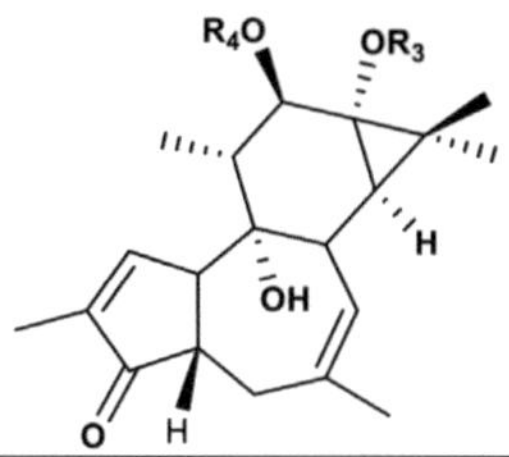

Compound	R1	R2	R3	R4
140	βH	βOH	isobutyroyl	isobutyroyl
141	βH	βOH	isobutyroyl	benzoyl
142	βH	ξOAc	isobutyroyl	benzoyl
143	βH	ξOH	isobutyroyl	isobutyroyl
144	βH	ξOH	isobutyroyl	benzoyl
195	βH	H	Ac	N-methylaminobenzoyl
196	αH	H	Ac	N-methylaminobenzoyl
197	αH	βOH	Ac	N-methylaminobenzoyl
198	βH	βOH	Ac	N-methylaminobenzoyl
199	βH	βOH	Ac	2,4,6-tetradecanoyl
200	βH	βOH	Ac	tetradecanoyl

Figure 14.
*Chemical structures of 4,20-dideoxyphorbol and 4,20-dideoxy-(4α)-phorbol derivatives (**140–144, 195–200**).*

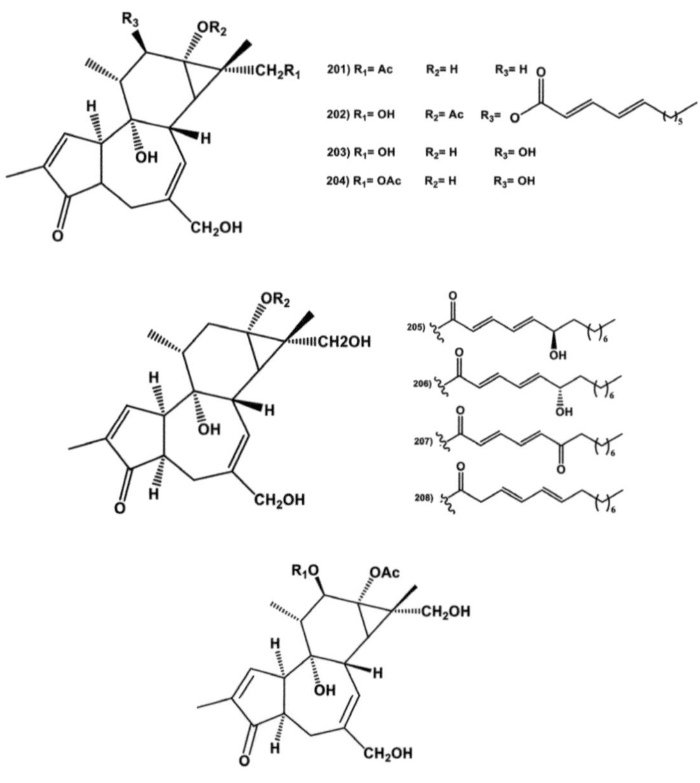

209) R1= hexanoyl
210) R1= dodecanoyl

Figure 15.
*Chemical structure of phorbol esters with 16 or 17 substitutions (**201–210**).*

DOI: http://dx.doi.org/10.5772/intechopen.1010654

2. Conclusions

The Euphorbiaceae family is primarily found in the intertropical zone, which is the largest geoastronomical zone on Earth. These plants are recognized worldwide for their toxic and medicinal properties and are notable for producing a milky liquid substance. Their high biological diversity is linked to their extensive chemical diversity, with diterpenes being a prominent class of secondary metabolites. Among these, "polycyclic" and "macrocyclic" diterpenes are particularly significant and have been used as chemotaxonomic markers for the family.

Phorbol esters (tigliano diterpenes) within this family are mainly confined to the genera *Croton*, *Euphorbia*, *Sapium*, and *Jatropha*. Within the group of diterpenes characteristic of the three genera described in the present work, it can be deduced that the genera *Euphorbia* and *Croton* are those that possess the highest number of tigliane derivatives, with the genus *Croton* being richer in phorbol derivatives, while the genus *Euphorbia* mainly biosynthesizes esters derived from 12-deoxyphorbol.

It is important to highlight that many compounds from this family, particularly those derived from the three genera mentioned above, have not yet undergone biological testing. Most existing studies have primarily concentrated on *in vitro* cytotoxic assays. There is still much to learn about the mechanisms of action and the structure-activity relationship, both of which are essential for understanding their potential as phytopharmaceuticals. However, the increasing interest in chemistry and pharmaceutical applications of species in this family may lead to significant advancements in the discovery and development of new compounds.

Acknowledgements

The author expresses gratitude to the Biology Program at FES-Zaragoza, UNAM.

Author details

Arturo Cano-Flores
Laboratorio de Biotransformaciones y Química de los Productos Naturales, Facultad de Estudios Superiores Zaragoza, UNAM, Ciudad de México, México

*Address all correspondence to: arturo.cano@zaragoza.unam.mx

References

[1] Ernst M, Grace OM, Saslis-Lagoudakis CH, Nilsson N, Simonsen HT, Ronsted N. Global medicinal uses of *Euphorbia* L. (*Euphorbiaceae*). Journal of Ethnopharmacology. 2015;**176**:90-101. DOI: 10.1016/j.jep.2015.10.025

[2] Prakash E, Gupta DK. Cytotoxic activities of extracts of medicinal plants of *Euphorbiacae* family studied on seven human cancer cell lines. Universal Journal of Plant Science. 2013;**1**:113-117. DOI: 10.13189/ujps.2013.010401

[3] Beutler JA, Alvarado AB, McCloud TG, Cragg GM. Distribution of phorbol ester bioactivity in the *Euphorbiaceae*. Phytotherapy Research. 1989;**3**:188-192. DOI: 10.1002/ptr.2650030507

[4] Hohmann J, Molnár J. *Euphorbiaceae* diterpenes: Plant toxins or promising molecules for the therapy? Acta Pharmaceutica Hungarica. 2004;**74**:149-157

[5] Clemens MJ, Trayner I, Menaya J. The role of protein kinase C isoenzymes in the regulation of cell proliferation and differentiation. Journal of Cell Science. 1992;**103**:881-887. DOI: 10.1242/jcs.103.4.881

[6] Goel G, Makkar HPS, Francis G, Becker K. Phorbol esters: Structure, biological activity, and toxicity in animals. International Journal of Toxicology. 2007;**26**:279-288. DOI: 10.1080/10915810701464641

[7] Vasas A, Hohmann J. *Euphorbia* diterpenes: Isolation, structure, biological activity synthesis (2008-2012). Chemical Reviews. 2014;**114**:8579-8612. DOI: 10.1021/cr400541j

[8] Xu Y, Tang P, Zhu M, Wang Y, Sun D, Li H, et al. Diterpenoids from the genus *Euphorbia*: Structure and biological activity (2013-2019). Phytochemistry. 2021;**190**:112846. DOI: 10.1016/j.phytochem.2021.112846

[9] Wang H-B, Wang X-Y, Liu L-P, Qin G-W, Kang T-G. Tigliane diterpenoids from the *Euphorbiaceae* and *Thymelaeaceae* families. Chemical Reviews. 2015;**115**:2975-3011. DOI: 10.1021/cr200397n

[10] Appendino G, Gaeta S. Tigliane diterpenoids. In: Kinghorn AD, Falk H, Gibbons S, Asakawa Y, Liu JK, Dirsch VM, editors. Progress in the Chemistry of Organic Natural Products. Vol. 125. Cham: Springer; 2024. DOI: 10.1007/978-3-031-67180-7

[11] Otsuki K, Li W. Tigliane and daphnane diterpenoids from *Thymelaeaceae* family: Chemistry, biological activity, and potential in drug discovery. Journal of Natural Medicine. 2023;**77**:625-643. DOI: 10.1007/s11418-023-01713-x

[12] Wender PA, Gentry ZO, Fanelli DJ, Luu-Nguyen QH, McAteer OD, Njoo E. Practical synthesis of the therapeutic leads tigilanol tiglate and its analogues. Nature Chemistry. 2022;**14**:1421-1426. DOI: 10.1038/s41557-022-01048-2

[13] Watanabe A, Nagatorno M, Hirose A, Hikone Y, Kishimoto N, Miura S, et al. Total syntheses of phorbol and 11 tigliane diterpenoids and their evaluation as HIV latency-reversing agents. Journal of the American Chemical Society. 2024;**146**:8746-8756. DOI: 10.1021/jacs.4c01589

[14] Miana GA, Riaz M, Shahzad-ul-Hussan S, Paracha RZ, Paracha UZ.

Prostratin: An overview. Mini Reviews in Medicinal Chemistry. 2015;**15**:1122-1130. DOI: 10.2174/1389557515666150511154108

[15] Jiaming Zhang JL, Hongdi L, Juan H, Shihong L. Esterification with a long-chain fatty acid elevates the exposure toxicity of tigliane diterpenoids from *Euphorbia fischeriana* roots against nematodes. Journal of Agricultural and Food Chemistry. 2023;**71**:12730-12740. DOI: 10.1021/acs.jafc.3c03460

[16] Yuan FY, Tang ZY, Huang D, Li W, Wu SQ, Huang JL, et al. Tigliane and rhamnofolane glycosides from *Euphorbia wallichii* prevent oxidative stress-induced neuronal death in PC-12 cells. Bioorganic Chemistry. 2022;**128**:106103. DOI: 10.1016/j.bioorg.2022.106103

[17] Xu WH, Liu WY, Liang Q. Chemical constituents from *Croton* species and their biological activities. Molecules. 2018;**12**:2333. DOI: 10.3390/molecules23092333

[18] Mfotie Nyoya E, Eloff JN, McGaw LJ. *Croton gratissimus* leaf extracts inhibit cancer cell growth by inducing caspase 3/7 activation with additional anti-inflammatory and antioxidant activities. BMC Complementary Alternative Medicine. 2018;**18**:305. DOI: 10.1186/s12906-018-2372-9

[19] Nath R, Roy S, De B, Choudhury MD. Anticancer and antioxidant activity of *Croton*: A review. International Journal of Pharmacy and Pharmaceutical Sciences. 2013;5:63-70. ISSN-0975-1491

[20] Terefe EM, Okalebo FA, Derese S, Langat MK, Mas-Claret E, Aljarba NH, et al. *In vitro* anti-HIV and cytotoxic effects of pure compounds isolated from *Croton macrostachyus* Hochst. Ex Delile. BMC Complementary Medicine and Therapies. 2022;**22**:159. DOI: 10.1186/s12906-022-03638-6

[21] Terefe EM, Okalebo FA, Derese S, Batiha GE, Youssef A, Alorabi M, et al. Cytotoxicity and anti-HIV activities of extracts of the twigs of *Croton dichogamus* Pax. BMC Complementary Medicine and Therapies. 2022;**22**:49. DOI: 10.1186/s12906-022-03532-1

[22] Terefe EM, Okalebo FA, Derese S, Muriuki J, Mas-Claret E, Langat MK. Anti-HIV crotocascarin ω from Kenyan *Croton dichogamus*. Natural Products Research. 2023;**37**:2809-2816. DOI: 10.1080/14786419.2022.2134998

[23] Salatino A, Faria SML, Negri G. Traditional uses, chemistry and pharmacology of *Croton* species (*Euphorbiaceae*). Journal of the Brazilian Chemical Society. 2007;**18**:11-33. DOI: 10.1590/S0103-50532007000100002

[24] Pieters LAC, Vlietinck AJ. ^{13}C NMR spectroscopy of phorbol esters. Magnetic Resonance in Chemistry. 1987;**25**:368-370. DOI: 10.1002/mrc.1260250420

[25] El-Mekkawy S, Meselhy MR, Nakamura N, Hattori M, Kawahata T, Otake T. Anti-HIV-1 phorbol esters from the seeds of *Croton tiglium*. Phytochemistry. 2000;**53**:457-464. DOI: 10.1016/s0031-9422(99)00556-7

[26] Marshall GT, Klocke JA, Lin LJ, Kinghorn D. Effects of diterpene esters of tigliane, daphnane, ingenane, and lathyrane types on pink bollworm, *Pectinophora gossypiella* saunders (Lepidoptera: Gelechiidae). Journal of Chemical Ecology. 1985;**11**:191-206. DOI: 10.1007/BF00988202

[27] Zhang XL, Wang L, Li F, Yu K, Wang MK. Cytotoxic phorbol esters of *Croton tiglium*. Journal of Natural Products. 2013;**76**:858-864. DOI: 10.1021/np300832n

[28] Marshall GT, Kinghorn AD. Short-chain phorbol ester constituents of

Croton oil. Journal of the American Oil Chemists' Society. 1984;**61**:1220-1225. DOI: 10.1007/BF02636256

[29] Rajiv RU, Erich H. A new cryptic irritant and cocarcinogen from seeds of *Croton sparciflorus*. Phytochemistry. 1976;**15**:1070-1072. DOI: 10.1016/S0031-9422(00)84408-8

[30] Hecker E, Schmidt R. Phorbolesters: The irritants and cocarcinogens of *Croton tiglium* L. Progress in the Chemistry of Organic Natural Products. 1974;**31**:377-467. DOI: 10.1007/978-3-7091-7094-6_7

[31] Kupchan SM, Uchida I, Branfman AR, Dailey RG, Fei BY. Antileukemic principles isolated from *Euphorbiaceae* plants. Science. 1976;**191**:571-572. DOI: 10.1126/science.1251193

[32] Marshall GT, Lee-Juian L, Kinghorn AD. Unambiguous ^{1}H-NMR assignment of phorbol triacetate using a two-dimensional NMR technique. Journal of Natural Products. 1985;**48**:823-825. DOI: 10.1021/np50041a021

[33] Ren FX, Ren FZ, Yang Y, Yu NJ, Zhang Y, Zhao YM. Tigliane diterpene esters from the leaves of *Croton tiglium*. Helvetica Chimica Acta. 2014;**97**:1014-1019. DOI: 10.1002/hlca.201300408

[34] Corlay N, Delang L, Girard-Valenciennes E, Neyts J, Clerc P, Smadja J, et al. Tigliane diterpenes from *Croton mauritianus* as inhibitors of chikungunya virus replication. Fitoterapia. 2014;**97**:87-91. DOI: 10.1016/j.fitote.2014.05.015

[35] Rios MY, Aguilar-Guadarrama AB. Nitrogen-containing phorbol esters from *Croton ciliatoglanduliferand* their effects on Cyclooxygenases-1 and -2. Journal of Natural Products. 2006;**69**:887-890. DOI: 10.1021/np0504311

[36] Ndunda B, Langat M, Wanjohi J, Midiwo J, Kerubo L. Alienusolin, a new 4α-deoxyphorbol ester derivative, and crotonimide C, a new glutarimide alkaloid from the Kenyan *Croton alienus*. Planta Medica. 2013;**79**:1762-1766. DOI: 10.1055/s-0033-1351044

[37] Benjamaa R, Moujanni A, Kaushik N, Choi EH, Essamadi AK, Kaushik NK. *Euphorbia* species latex: A comprehensive review on phytochemistry and biological activities. Frontiers in Plant Science. 2022;**13**:1008881. DOI: 10.3389/fpls.2022.1008881

[38] Kemboi D, Peter X, Langat M, Tembu J. A review of the ethnomedicinal uses, biological activities, and triterpenoids of *Euphorbia* species. Molecules. 2020;**25**:4019. DOI: 10.3390/molecules25174019

[39] Rojas-JiménezS, Valladares-CisnerosMG, Salinas-Sánchez DO, Pérez-Ramos J, Sánchez-Pérez L, Pérez-Gutiérrez S, et al. Anti-inflammatory and cytotoxic compounds isolated from plants of *Euphorbia* genus. Molecules. 2024;**29**:1083. DOI: 10.3390/molecules29051083

[40] Huang Y-S, Lu Y, Chen C-H, Lee K-H, Chen DF. Potent anti-VIH ingenane diterpenoids from *Euphorbia ebracteolata*. Journal of Natural Products. 2019;**82**:1587-1592. DOI: 10.1021/acs.jnatprod.9b00088

[41] Salehi B, Iriti M, Vitalini S, Antolak H, Pawlikowska E, Kręgiel D, et al. *Euphorbia*-derived natural products with potential for use in health maintenance. Biomolecules. 2019;**9**:337. DOI: 10.3390/biom9080337

[42] Shu-Qi W, Xinying Z, Tao Y, Fang-Yu Y, Shiyou Z, Dong H, et al. Discovery of ingenane diterpenoids from *Euphorbia hylonoma* as antiadipogenic

agents. Journal of Natural Products. 2023;**86**:2691-2702. DOI: 10.1021/acs.jnatprod.3c00822

[43] Benmerache A, Labed A, Kabouche A, Kabouche Z, Benmerache A, Alabdul Magid A, et al. Isolation and characterisation of cytotoxic compounds from *Euphorbia clementei* Boiss. Natural Products Research. 2017;**31**:2091-2098. DOI: 10.1080/14786419.2016.1277345

[44] Lanzotti V, Barile E, Scambia G, Ferlini C. Cyparissins A and B, jatrophane diterpenes from *Euphorbia cyparissias* as Pgp inhibitors and cytotoxic agents against ovarian cancer cell lines. Fitoterapia. 2015;**104**:75-79. DOI: 10.1016/j.fitote.2015.05.012

[45] Yuan WJ, Yang GP, Zhang JH, Zhang Y, Chen DZ, Li SL, et al. Three new diterpenes with cytotoxic activity from the roots of *Euphorbia ebracteolate* Hayata. Phytochemistry Letters. 2016;**18**:17179. DOI: 10.1016/j.phytol.2026.10.008

[46] Maru N, Chikaraishi N, Yokota K, Kawazoe Y, Uemura D, Jolkinolide F. A cytotoxic diterpenoid from *Euphorbia jolkinii*. Chemistry Letters. 2013;**42**:756-757. DOI: 10.1246/cl.130280

[47] Dong M, Chen XQ, Chen CH, Li RT. Terpenes from *Euphorbia antiquorum* and their in vitro anti-HIV activity. Chemistry & Biodiversity. 2018;**15**:e1700560. DOI: 10.1002/cbdv.201700560

[48] Yan SL, Li YH, Chen XQ, Liu D, Chen CH, Li RT. Diterpenes from the stem bark of *Euphorbia neriifolia* and their in vitro anti-HIV activity. Phytochemistry. 2018;**145**:40-47. DOI: 10.1016/j.phytochem.2017.10.006

[49] Zhang CY, Wu YL, Zhang P, Chen ZZ, Li H, Chen LX. Anti-inflammatory lathyrane diterpenoids from *Euphorbia lathyris*. Journal of Natural Products. 2019;**82**:756-764. DOI: 10.1021/acs.jnatprod.8b00600

[50] Zhang JS, Weng HZ, Huang JL, Tang GH, Yin S. Anti-inflammatory ingenane diterpenoids from the roots of *Euphorbia kansui*. Planta Medica. 2018;**84**:1334-1339. DOI: 10.1055/a-0646-4306

[51] Xu J, Kang J, Cao X, Sun X, Yu S, Zhang X, et al. Characterization of diterpenes from *Euphorbia prolifera* and their antifungal activities against phytopathogenic fungi. Journal of Agricultural and Food Chemistry. 2015;**63**:5902-5910. DOI: 10.1021/acs.jafc.5b02021

[52] Kusz N, Orvos P, Bereczki L, Fertey P, Bombicz P, Csorba A, et al. Diterpenoids from *Euphorbia dulcis* with potassium ion channel inhibitory activity with selective G protein-activated inwardly rectifying ion channel (GIRK) blocking effect. Journal of Natural Products. 2018;**81**:2483-2492. DOI: 10.1021/acs.jnatprod.8b00500

[53] Kusz N, Orvos P, Csorba A, Talosi L, Chaieb M, Hohmann J, et al. Jatrophane diterpenes from *Euphorbia guyoniana* are new potent inhibitors of atrial GIRK channels. Tetrahedron. 2016;**72**:5724-5728. DOI: 10.1016/j.tet.2016.08.007

[54] Wei YL, Yu ZL, Huo XK, Tian XG, Feng L, Huang SS, et al. Diterpenoids from the roots of *Euphorbia fischeriana* and their inhibitory effects on α-glucosidase. Journal of Asian Natural Products Research. 2018;**20**:977-984. DOI: 10.1080/10286020.2017.1367923

[55] Huang CS, Luo SH, Li YL, Li CH, Hua J, Liu Y, et al. Antifeedant and antiviral diterpenoids from the fresh roots of *Euphorbia jolkinii*. Natural

Products and Bioprospecting. 2014;**4**:91-100. DOI: 10.1007/s13659-014-0009-3

[56] Krstic G, Jadranin M, Todorovic NM, Pesic M, Stankovic T, Aljancic IS, et al. Jatrophane diterpenoids with multidrug resistance modulating activity from the latex of *Euphorbia nicaeensis*. Phytochemistry. 2018;**148**:104-112. DOI: 10.1016/j.phytochem.2018.01.016

[57] Gao J, Aisa HA. Terpenoids from *Euphorbia soongarica* and their multidrug resistance reversal activity. Journal of Natural Products. 2017;**80**:1767-1775. DOI: 10.1021/acs.jnatprod.6b01099

[58] Wang P, Xie C, An L, Yang X, Xi Y, Yuan S, et al. Bioactive diterpenoids from the stems of *Euphorbia royleana*. Journal of Natural Products. 2019;**82**:183-193. DOI: 10.1021/acs.jnatprod.8b00493

[59] Ferreira RJ, Kincses A, Gajdács M, Spengler G, dos Santos DJVA, Molnár J, et al. Terpenoids from *Euphorbia pedroi* as multidrug-resistance reversers. Journal of Natural Products. 2018;**81**:2032-2040. DOI: 10.1021/acs.jnatprod.8b00326

[60] Islam MN, Pramanik MEA, Hossain MA. Identification of leading compounds from *Euphorbia neriifolia* (Dudsor) extracts as a potential inhibitor of SARS-CoV-2 ACE2-RBDS1 receptor complex: An insight from molecular docking ADMET profiling and MD-simulation studies. Euroasian Journal of Hepato-Gastroenterology. 2023;**13**:89-107. DOI: 10.5005/jp-journals-10018-1414

[61] Wang HB, Chu WJ, Wang-Y JP, Wang YB, Yu Q, Qin GW. Diterpenoids from the roots of *Euphorbia fischeriana*. Journal of Asian Natural Products Research. 2010;**12**:1038-1043. DOI: 10.1080/10286020.2010.532490

[62] Fürstenberger G, Hecker E. On the active principles of the Euphorbiaceae, XII. Highly unsaturated irritant diterpene esters from *Euphorbia tirucalli* originating from Madagascar. Journal of Natural Products. 1986;**49**:386-397. DOI: 10.1021/np50045a002

[63] Evans FJ. A new phorbol triester from the latices of *Euphorbia frankiana* and *E. Coerulescens*. Phytochemistry. 1977;**16**:395-396. DOI: 10.1016/00319422(77)80075-7

[64] Urones JG, Barcala PB, Cuadrado MJS, Marcos IS. Diterpenes from the latex of *Euphorbia broteri*. Phytochemistry. 1988;**27**:207-212. DOI: 10.1016/0031-9422(88)80615-0

[65] Pan LL, Fang PL, Zhang XJ, Ni W, Li L, Yang LM, et al. Tigliane-type diterpenoid glycosides from *Euphorbia fischeriana*. Journal of Natural Products. 2011;**74**:1508-1512. DOI: 10.1021/np200058c

[66] Zayed S, Sorg B, Hecker E. Structure activity relations of polyfunctional diterpenes of the tigliane type, VI. Planta Medica. 1984;**50**:65-69. DOI: 10.1055/s-2007-969623

[67] Ma Q-G, Liu W-Z, Xiao-Yun W, Zhou T-X, Qin G-W. Diterpenoids from *Euphorbia fischeriana*. Phytochemistry. 1997;**44**:663-666. DOI: 10.1016/s0031-9422(96)00605-x

[68] Sosath S, Ott HH, Hecker E. Irritant principles of the spurge family (*Euphorbiaceae*). XIII. Oligocyclic and macrocyclic diterpene esters from latices of some *Euphorbia* species utilized as source plants of honey. Journal of Natural Products. 1988;**51**:1062-1074. DOI: 10.1021/np50060a003

[69] Baloch IB, Baloch MK, Saqib QN. Anti-tumor 12-deoxyphorbol esters from

Euphorbia cornigera. Europan of Journal Medicinal Chemistry. 2008;**43**:274-281. DOI: 10.1016/j.ejmech.2007.03.016

[70] Hergenhahn M, Kusumoto S, Hecker E. On the active principles of the spurge family (Euphorbiaceae). Journal Cancer Research and Clinical Oncology. 1984;**108**:98-109. DOI: 10.1007/BF00390980

[71] Evans FJ, Schmidt RJ. Two new toxins from the latex of *Euphorbia poisonii*. Phytochemistry. 1976;**15**:333-335. DOI: 10.1016/s0031-9422(00)89024-x

[72] Evans FJ, Kinghorn AD, Schmidt RJ. Some naturally occurring skin irritants. Acta Pharmacologica et Toxicologica. 1975;**37**:250-256. DOI: 10.1111/j.1600-0773.1975.tb00842.x

[73] Fatope MO, Zeng L, Ohayaga JE, Shi G, McLaughlin JL. Selectively cytotoxic diterpenes from *Euphorbia poisonii*. Journal of Medical Chemistry. 1996;**16**:1005-1008. DOI: 10.1021/jm950509i

[74] Wang YB, Huang R, Wang HB, Jin HZ, Lou LG, Qin GW. Diterpenoids from the roots of *Euphorbia fischeriana*. Journal of Natural Products. 2006;**69**:967-970. DOI: 10.1021/np0600088

[75] Yang Y, Chen X, Luan F, Wang M, Wang Z, Wang J, et al. *Euphorbia helioscopia* L.: A phytochemical and pharmacological overview. Phytochemistry. 2021;**184**:112649. DOI: 10.1016/j.phytochem.2020.112649

[76] Schmidt RJ, Evans FJ. Two minor diterpenes from *Euphorbia* latex. Phytochemistry. 1978;**17**:1436-1437. DOI: 10.1016/s0031-9422(00)94608-9

[77] Evans FJ, Kinghorn AD. New diesters of 12-deoxy-phorbol. Phytochemistry. 1975;**14**:1669-1670. DOI: 10.1016/0031-9422(75)85388-x

[78] Fürstenberger VG, Hecker E. Zum Wirkungsmechanismus cocarcinogener Pflanzeninhaltsstoffe. Planta Medica. 1972;**22**:241-266. DOI: 10.1055/s-0028-1099610

[79] Mali PY, Panchal SS. *Euphorbia tirucalli* L.: Review on morphology, medicinal uses, phytochemistry and pharmacological activities. Asian Pacific Journal of Tropical Biomedicine. 2017;**7**:603-613. DOI: 10.1016/j.apjtb.2017.06.002

[80] Fürstenberger G, Hecker E. The new diterpene 4-deoxyphorbol and its highly unsaturated irritant diesters. Tetrahedron Letters. 1977;**18**:925-928. DOI: 10.1016/s0040-4039(01)92793-5

[81] Marco JA, Sanz-Cervera JF, Checa J, Palomares E, Fraga BM. Jatrophane and tigliane diterpenes from the latex of *Euphorbia obtusifolia*. Phytochemistry. 1999;**52**:479-485. DOI: 10.1016/S0031-9422(99)00166-1

[82] Betancur-Galvis L, Palomares E, Marco JA, Estornell E. Tigliane diterpenes from the latex of *Euphorbia obtusifolia* with inhibitory activity on the mammalian mitochondrial respiratory chain. Journal of Ethnopharmacology. 2003;**85**:279-282. DOI: 10.1016/S0378-847(03)0010-2

[83] Dagang W, Sorg B, Hecker E. Oligo- and macrocyclic diterpenes in *Thymelaeaceae* and Euphorbiaceae occurring and utilized in Yunnan (Southwest China). 6. Tigliane type diterpene esters from latex of *Euphorbia prolifera*. Phytotherapy Research. 1994;**8**:95-99. DOI: 10.1002/ptr.2650080209

[84] Zayed SM, Farghaly M, Soliman SM, Gotta H, Sorg B, Hecker E. Dietary

cancer risk from conditional carcinogens (tumor promoters) in produce of livestock fed on species of spurge (*Euphorbiaceae*). Journal of Cancer Research Oncology. 2001;**127**:40-47. DOI: 10.1007/s004320000214

[85] Appendino G, Belloro E, Tron GC, Jakupovic J, Ballero M. Diterpenoids from *Euphorbia pithyusa* subsp. *cupanii*. Journal of Natural Products. 1999;**62**:1399-1404. DOI: 10.1021/np990209u

[86] Sulyok E, Vasas A, Rédei D, Dombi G, Hohmann J. Isolation and structure determination of new 4,12-dideoxyphorbol esters from *Euphorbia pannonica* Host. Tetrahedron. 2009;**65**:4013-4016. DOI: 1016/j.tet.2009.03.014

[87] Haba H, Lavaud C, Harkat H, Alabdul Magid A, Marcourt L, Benkhaled M. Diterpenoids and triterpenoids from *Euphorbia guyoniana*. Phytochemistry. 2007;**68**:1255-1260. DOI: 10.1016/j.phytochem.2007.01.017

[88] Duarte N, Lage H, Ferreira MJ. Three new jatrophane polyesters and antiproliferative constituents from *Euphorbia tuckeyana*. Planta Medica. 2008;**74**:61-68. DOI: 10.1055/s-2007-993765

[89] Xu J, Jin D, Guo P, Xie C, Fang L, Guo Y. *Three new* myrsinol diterpenes from *Euphorbia prolifera* and their neuroprotective activities. Molecules. 2012;**17**:9520-9528. DOI: 10.3390/molecules17089520

[90] Hergenhahn M, Adolf W, Hecker E. Resiniferatoxin and other esters of novel polyfuncticnal diterpenes from *Euphorbia resinifera* and *unispina*. Tetrahedron Letters. 1975;**16**:1595-1598. DOI: 10.1016/s0040-4039(00)72207-6

[91] Ezzanad A, De los Reyes C, Macías-Sánchez AJ, Hernández-Galán R. Isolation and identification of 12-deoxyphorbol esters from *Euphorbia resinifera* berg latex: Targeted and biased non-targeted identification of 12-deoxyphorbol esters by UHPLC-HRMS^{E}. Plants. 2023;**12**:3846. DOI: 10.3390/plants12223846

[92] Forgo P, Rédei D, Hajdu Z, Szabó P, Szabó L, Hohmann J. Unusual tigliane diterpenes from *Euphorbia grandicornis*. Journal of Natural Products. 2011;**74**:639-643. DOI: 10.1021/np100673s

[93] Evans FJ, Schmidt RJ. The succulent euphorbias of Nigeria. III. Structure and potency of the aromatic ester diterpenes of *Euphorbia poissonii* Pax. Acta Pharmacologica et Toxicologica. 1979;**45**:181-191. DOI: 10.1111/j.1600-0773.1979.tb02380.x

[94] Gschwendt M, Hecker E. Tumor-promoting compounds from *Euphorbia cooperi* di- and triesters of 16-hydroxy-12-desoxy-phorbol. Tetrahedron Letters. 1970;**11**:567-570. DOI: 10.1016/s0040-4039(01)97771-8

[95] Popplewell WL, Marais EA, Brand L, Harvey BH, Davies-Coleman MT. Euphorbias of South Africa: Two new phorbol esters from *Euphorbia bothae*. South African Journal of Chemistry. 2010;**63**:175-179. Available from: https://hdl.handle.net/10520/EJC24500

[96] Al Muqarrabun LMR, Ahmat N, Aris SRS. A review of the medicinal uses, phytochemistry and pharmacology of the genus *Sapium*. Journal of Ethnopharmacology. 2014;**155**:9-20. DOI: 10.1016/j.jep.2014.05.028

[97] He Q, Zhang L, Li T, Li C, Song H, Fan P. *Genus Sapium (Euphorbiaceae):* A review on traditional uses,

phytochemistry, and pharmacology. Journal of Ethnopharmacology. 2021;**277**:114206. DOI: 10.1016/j.jep.2021.114206

[98] Ahmed Y, Sohrab MH, Al-Reza SM, Tareq FS, Hasan CM, Sattar MA. Antimicrobial and cytotoxic constituents from leaves of *Sapium baccatum*. Food Chemical Toxicology. 2010;**48**:549-552. DOI: 10.1016/j.fct.2009.11.030

[99] Fu R, Zhang YT, Guo YR, Huang QL, Peng T, Xu Y, et al. Antioxidant and anti-inflammatory activities of the phenolic extracts of *Sapium sebiferum* (L.) Roxb. leaves. Journal of Ethnopharmacology. 2013;**147**:517-524. DOI: 10.1016/j.jep.2013.03.058

[100] Woldemichael GM, Lugo MTG, Franzblau SG, Wang Y, Suarez E, Timmermann BN. *Mycobacterium tuberculosis* growth inhibition by constituents of *Sapium haematospermum*. Journal of Natural Products. 2004;**67**:598-603. DOI: 10.1021/np0303411

[101] Taylor SE, Evans FJ, Gafur MA, Choudhury AK. Sapentoxin D: A new phorbol ester from *Sapium indicum*. Journal of Natural Products. 1981;**44**:729-731. DOI: 10.1021/np50018a021

[102] Brooks G, Morrice NA, Ellis C, Aitken A, Evans AT, Evans FJ. Toxic phorbol esters from Chinese tallow stimulate protein kinase C. Toxicon. 1987;**25**:1229-1233. DOI: 10.1016/0041-0101(87)90141-3

[103] Seip EH, Ott H, Hecker E. Skin irritant and tumor promoting diterpene esters of the tigliane type from the Chinese tallow tree (*Sapium sebiferum*). Planta Medica. 1983;**49**:199-203. DOI: 10.1055/s-2007-969850

[104] Grosvenor SNJ, McLean S, Reynolds WF, Tinto WF. Two phorbol esters from *Sapium hippomane*. Natural Product Communications. 2007;**2**. DOI: 10.1177/1934578XO700200203

[105] Chumkaew P, Karalai C, Ponglimanont C, Chantrapromma K. Antimycobacterial activity of phorbol esters from the fruits of *Sapium indicum*. Journal of Natural Products. 2003;**66**:540-543. DOI: 10.1021/np0204489

[106] Edwards MC, Taylor SE, Williamson EM, Evans FJ. New phorbol and deoxyphorbol esters: Isolation and relative potencies in inducing platelet aggregation and erythema of skin. Acta Pharmacologica et Toxicologica. 1983;**53**:177-187. DOI: 10.1111/j.1600-0773.1983.tb01122.x

[107] Taylor SE, Gafur MA, Choudhury AK, Evans FJ. 4-desoxyforbol y 4α-desoxyforbol aldehydes: New diterpenes and their esteres. Tetrahedron Letters. 1981;**22**:3321-3324. DOI: 10.1016/S00040-4039(01)81895.5

[108] Taylor SE, Gafur MA, Choudhury AK, Evans FJ, Sapintoxin A. A new biologically active nitrogen containing phorbol ester. Experientia. 1981;**37**:681-682. DOI: 10.1007/bf01967918

[109] Taylor SE, Gafur MA, Choudhury AK, Evans FJ. Sapatoxins, aliphatic ester tigliane diterpenes from *Sapium indicum*. Phytochemistry. 1982;**21**:405-407. DOI: 10.1016/s0031-9422(00)95276-2

[110] Taylor SE, Williamson EM, Evans FJ. Phorbol derivatives from *Sapium insigne*. Phytochemistry. 1983;**22**:1231-1233. DOI: 10.1016/0031-9422(83)80228-3

[111] Taylor SE, Gafur MA, Choudhury AK, Evans FJ.

Nitrogen-containing phorbol derivatives of *Sapium indicum*. Phytochemistry. 1981;**20**:2749-2751. DOI: 10.1016/0031-9422(81)85279-x

[112] Liu HB, Zhang H, Yu JH, Xu CH, Ding J, Yue JM. Cytotoxic diterpenoids from *Sapium insigne*. Journal of Natural Products. 2012;**75**:722-727. DOI: 10.1021/np300004y

[113] Steinmann VW. Diversity and endemism of the Euphorbiaceae family in Mexico. Acta Botanica Mexicana. 2002;**61**:61-93. DOI: 10.21829/abm61.2002.909

[114] Devkota HP, Basnet P, Yahara S. Diterpene esters and phenolic compounds from *Sapium insigne* (ROYLE) BENTH. ex HOOK. fil. Chemical Pharmaceutical Bulletin. 2009;**57**:1289-1291. DOI: 10.1248/cpb.57.1289

[115] Ohigashi H, Ohtuka T, Hirota M, Kosmimizu K, Tokuda H, Ito Y. Tigliane type diterpene esters with Epstein-Barr virus-inducing activity *Sapium sebiferum*. Agricultural and Biological Chemistry. 1983;**47**:1617-1622. DOI: 10.1080/00021369.1983.10865799

Section 2

Applications

Chapter 4

Cyanobolites: Future Prospects in Therapeutics and Biological Innovation

Kundan Ojha, Satish Dubey, Akanksha Singh, Ashwini Kumar Dixit and Priyanka Kumari

Abstract

Cyanobacteria, a group of photosynthetic prokaryotes, are gaining significant attention for their ability to produce bioactive secondary metabolites, commonly referred to as cyanobolites. These compounds, often synthesized in response to environmental stressors, exhibit a broad spectrum of biological activities, including antimicrobial, antiviral, antifungal, and anticancer properties. Recent advances in isolation techniques and molecular characterization have accelerated the identification of these structurally unique cyanobolites, many of which hold promising therapeutic potential. Cyanobolites have emerged as a prolific source of novel bioactive compounds with selective cytotoxicity against cancer cells and inhibitory effects on various microbial pathogens. This chapter explores the current landscape of cyanobolites, emphasizing their emerging roles in drug discovery, biotechnological innovation, and environmental sustainability. The discussion focuses on the discovery, bioactivity, molecular mechanisms, and potential future applications of cyanobolites in pharmaceutical development and biological sciences. The future prospects of cyanobolites indicate their expanding role in shaping the next generation of therapeutic agents and contributing to advancements in environmental and biomedical research. This chapter provides a comprehensive overview of cyanobolites, highlighting their significance in ushering in a new era of biological and pharmacological progress.

Keywords: cyanobacteria, cyanobolites, bioactivity, drug discovery, pharmacological potential

1. Introduction

Cyanobacteria are widely recognized as some of the earliest photosynthetic organisms on Earth. These ancient microorganisms played a pivotal role in transforming Earth's early atmosphere by releasing oxygen through photosynthesis, thereby enabling the evolution of aerobic life forms, including eukaryotic organisms [1]. Classified as gram-negative prokaryotes, cyanobacteria possess unique structural features that bridge the gap between bacteria and algae. Their cellular structure includes a primitive nucleus, thylakoid membranes organized in a minimal manner, hexagonal gas vesicles, and specialized inclusions such as carboxysomes, ribosomes, phycobilisomes, and storage granules comprising cyanophycin, glycogen, lipids, and polyphosphates [2, 3]. Cyanobacteria are distributed across a broad range of habitats, including saline waters, alkaline lakes, freshwater ecosystems, rice fields, cave surfaces, deserts, polar regions, geothermal springs, and marine environments. Their widespread ecological distribution reflects their remarkable adaptability and ability to thrive under diverse environmental conditions. Cyanobacteria have evolved various physiological and biochemical mechanisms that enable them to withstand extreme conditions, such as fluctuations in temperature, light intensity, pH, and nutrient availability, ensuring their survival and ecological success [4, 5].

Natural products have long been regarded as a rich source of potential therapeutic agents. Their significance in drug discovery and development arises from their structural diversity, which is often unattainable through conventional combinatorial chemistry [6]. This structural diversity is crucial for identifying novel lead compounds with low molecular weights. However, despite their potential, less than 10% of the Earth's biodiversity has been explored for biological activity, suggesting that numerous valuable natural products remain undiscovered. Secondary metabolites, a class of organic compounds, play a vital role in the long-term survival and ecological interactions of organisms. These compounds, synthesized by specific cells or tissues, exhibit significant variability in both type and quantity across individuals of the same species. This variability highlights their diverse functions in defense mechanisms, environmental adaptation, and ecological competition [7].

2. Introduction to cyanobolites

Cyanobacteria are prolific producers of secondary metabolites, creating a wide range of compounds in response to environmental challenges [8]. These metabolites encompass peptides, alkaloids, polyketides, terpenoids, and other UV-absorbing substances, each serving various protective functions such as predator evasion, sensory roles, light protection, and antioxidant properties. Notably, nonribosomal peptides (NRPs) and polyketides are key secondary metabolite classes in cyanobacteria. NRPs, synthesized by nonribosomal peptide synthases (NRPS), are characterized by complex structures incorporating diverse amino acids and modifications, often displaying antimicrobial, cytotoxic, and immunosuppressive properties [9]. Hybrid metabolites that combine features of polyketides and NRPs lead to unique chemical scaffolds with specialized bioactive properties. Microcystin-LR, produced through the interaction of NRPS and polyketide synthase (PKS) modules, has drawn attention for its potential anticancer activity due to its selective cytotoxicity toward tumor cells [10, 11]. Additionally, nodularin, another cyanobacterial metabolite, has been reported to exhibit similar bioactive potential [9].

Ribosomal peptides (RPs) such as cyanobactins are synthesized on ribosomes and subsequently undergo extensive post-translational modifications, yielding highly complex structures. Examples include cyclic peptides like patellamides, known for their antitumor and antiviral properties, and linear peptides such as aeruginosamides, which exhibit antibacterial activity [12]. In addition, bacteriocins produced by cyanobacteria show promising antimicrobial activities, further emphasizing their role in microbial competition and survival [13]. Alkaloids, particularly indole alkaloids, represent another important group of cyanobacterial secondary metabolites. These nitrogen-containing compounds are often biologically active and have been linked to various pharmacological properties. Notable examples include saxitoxin, a potent neurotoxin responsible for paralytic shellfish poisoning, and hapalindoles, which possess antibacterial and anticancer activities [14]. Moreover, hapalindole-type alkaloids have been highlighted for their diverse structural motifs and high potency against pathogenic microbes [14].

Terpenoids, another significant class of secondary metabolites, play important ecological roles, including allelopathy and antifungal activity. These include compounds, such as lyngbyatoxin A, a potent skin irritant, and tolytoxin, which exhibits antifungal properties [15]. Additionally, cyanobacterial carotenoids, which serve as photoprotective pigments, contribute to cellular protection against reactive oxygen species generated by high-light exposure. Cyanobacteria's ability to produce such a broad array of bioactive secondary metabolites makes them invaluable for biotechnological applications, including drug discovery, nutraceuticals, and environmental remediation. Despite their immense potential, many cyanobacterial metabolites remain unexplored, offering a vast reservoir for future research and therapeutic development [16].

3. Cyanobolites and environmental stress response

Cyanobacteria are among the most ancient and adaptable organisms on Earth, thriving in a wide range of environmental conditions—from freshwater lakes to extreme environments like deserts and hot springs. This ecological success can be attributed to their ability to produce a diverse array of bioactive compounds, collectively termed cyanobolites (secondary metabolites of cyanobacterial origin). Cyanobolites play a crucial role in the survival and competitive advantage of cyanobacteria by mediating various stress responses, such as defense against predators, adaptation to nutrient limitations, and protection from UV radiation and oxidative damage [17].

3.1 Types of cyanobolites in stress response

3.1.1 Peptides and polyketides

Among cyanobolites, nonribosomal peptides (NRPs) and polyketides are well-studied for their protective roles in environmental stress conditions. NRPs are synthesized by nonribosomal peptide synthases (NRPs), while polyketides are formed by polyketide synthases (PKSs). These compounds often function as toxins to deter grazers and predators, as well as antimicrobial agents to outcompete other microorganisms in shared habitats [17]. For example, microcystins—cyclic peptides produced by various cyanobacterial species—play a significant role in defense against zooplankton grazing. In addition, microcystins help cyanobacteria mitigate oxidative stress caused by high-light intensity or nutrient limitation by interacting with cellular antioxidant systems [18].

3.1.2 Alkaloids

Alkaloids are another important group of cyanobolites involved in stress response. Nitrogen-containing compounds such as hapalindoles and saxitoxins serve multiple ecological functions. Saxitoxins, for instance, are potent neurotoxins that protect cyanobacteria from predation by zooplankton and fish. Additionally, these alkaloids may provide protection against UV radiation by acting as UV-absorbing compounds [18]. Some alkaloids also function as signaling molecules, enabling cyanobacteria to sense environmental changes and modulate their physiological responses accordingly [19].

3.1.3 Terpenoids

Cyanobacterial terpenoids, including carotenoids and lyngbyatoxins, are known for their photoprotective and anti-stress properties. Carotenoids act as essential components of the photosynthetic machinery, protecting the cells from reactive oxygen species (ROS) generated during high-light exposure. Lyngbyatoxins, on the other hand, are toxic secondary metabolites that serve as deterrents against grazers and microbial competitors [20].

3.1.4 Scytonemin

One of the most prominent cyanobolites involved in UV protection is scytonemin, a yellow-brown pigment that accumulates in the extracellular sheath of cyanobacteria. Scytonemin acts as a natural sunscreen, absorbing UV radiation and preventing damage to cellular components. This compound is particularly important for cyanobacteria inhabiting extreme environments such as deserts and polar regions, where UV exposure is high [21].

3.2 Role in nutrient stress response

Cyanobolites are also critical for cyanobacteria in responding to nutrient limitations. Under nitrogen-depleted conditions, cyanobacteria produce specialized heterocyst cells for nitrogen fixation. Secondary metabolites such as cyanophycin, a nitrogen storage polymer, help cyanobacteria store excess nitrogen during periods of abundance, ensuring survival during scarcity. Additionally, polyphosphate granules are formed in response to phosphorus limitation, enabling cyanobacteria to manage and utilize phosphorus efficiently during growth cycles [22].

3.3 Oxidative stress management

Oxidative stress caused by reactive oxygen species (ROS) is another significant environmental challenge that cyanobacteria encounter. Cyanobolites such as glutathione, tocopherols, and certain polyketides act as antioxidants, scavenging free radicals and protecting cells from oxidative damage. Moreover, compounds like microcystins have been shown to modulate the activity of protein phosphatases, which helps in maintaining cellular homeostasis under stress conditions [23].

3.4 Ecological significance and biotechnological potential

The ability of cyanobacteria to produce such a wide range of stress-related metabolites has profound ecological implications. Cyanobolites not only enhance cyanobacterial survival but also shape microbial community dynamics by influencing competition and predator-prey interactions. Furthermore, due to their bioactive properties, many of these compounds hold significant promise for biotechnological applications, particularly in pharmaceuticals and environmental biotechnology [19]. Compounds such as microcystins, hapalindoles, and scytonemins are being explored for their potential as anticancer agents, antimicrobial drugs, and UV protectants, respectively [24].

4. Biological activities of cyanobolites

Over the past few years, several important classes of cytotoxic compounds derived from marine cyanobacteria have been identified, particularly from various strains of the genera *Lyngbya* or *Leptolyngbya*. These findings highlight the vast potential of marine cyanobacteria to integrate NRPS (nonribosomal peptide synthetase) and PKS (polyketide synthase) biosynthetic pathways. Many of these metabolites are considered to have significant potential for development as chemotherapeutic agents [25].

4.1 Somocystinamide A

One notable example is *Lyngbya majuscula*, which produces a highly unusual neurotoxic dimeric lipopeptide named somocystinamide A. This compound contains two unique N-methyl enamide groups. A subsequent screening program revealed that cancer cells with an active caspase-8 system are highly sensitive to the apoptosis-inducing effects of somocystinamide A. The compound has also been shown to inhibit neural tube formation in endothelial cells and interact with lipid rafts. It is proposed that somocystinamide A exerts its effects by activating a death-inducing signaling complex within the cell membrane. This mechanism sequentially activates caspase-8 and caspase-3, thereby triggering the extrinsic apoptosis pathway. These findings underscore the potential of cyanobacterial secondary metabolites in targeting cancer cells and disrupting critical cellular processes, offering new avenues for cancer therapy development [26].

4.2 Apratoxins

Apratoxins are a unique family of cyclic lipopeptide metabolites produced by cyanobacteria, recognized for their exceptional cytotoxicity against cancer cells. Structurally, they consist of two polyketide sections interspersed with amino acid residues, with a distinctive tertiary butyl group marking the initiation of biosynthesis. Seven apratoxins (A–G) have been identified, each displaying unique modifications such as the absence of N- or C-methyl groups (B, C, E, and G), the addition of an extra polyketide synthase (PKS) module (D), or the replacement of a terminal proline residue with N-methyl alanine (F and G) [27]. Apratoxins A and F, in particular, exhibit sub-nanomolar LD_{50} values in various cancer cell lines, making them the most potent variants. These compounds disrupt multiple cancer-related processes, including targeting Heat Shock Protein 90 (HSP90) and impairing the secretory pathway, which are critical for cancer cell survival and proliferation [28]. The biosynthesis of apratoxins highlights the ingenuity of cyanobacterial metabolic pathways, involving the integration of polyketide synthase (PKS) and nonribosomal peptide synthetase (NRPS) systems. The successful total synthesis of apratoxin A in laboratories has enabled structure-activity relationship (SAR) studies, facilitating the development of synthetic analogs for detailed pharmacological investigations. These studies reveal apratoxins' potential as therapeutic agents, disrupting cancer cell survival mechanisms through apoptosis induction and pathway interference. With their unique structural features and multifaceted mechanisms of action, apratoxins represent a promising frontier in the search for innovative cancer therapies [29].

4.3 Coibamide A

Coibamide A is a highly unique peptide derived from cyanobacteria, distinguished by its unconventional structure, which notably lacks typical polyketide synthase (PKS) elements commonly seen in cyanobacterial metabolites. Its architecture includes a large macrocyclic ring and a linear segment that terminates with an *N,N*-dimethyl valine at the amino terminus. A defining characteristic of coibamide A is its extensive *N*- and *O*-methylation, which contributes to its structural stability and bioactivity [30]. This compound demonstrates potent anti-proliferative effects on cancer cells grown *in vitro*, with activity observed at low nanomolar concentrations, making it a promising candidate for further anticancer research. However, its precise mechanism of action and molecular targets remain elusive, presenting a significant opportunity for future studies to uncover the biochemical pathways influenced by this compound. The structural novelty and cytotoxic potency of coibamide A highlight its potential as a lead compound for developing new cancer therapies [30].

4.4 Bisebromoamide

The total synthesis of bisebromoamide confirmed its planar structure and established the configuration of seven out of its eight chiral centers. However, this synthetic study also corrected the stereochemistry of a quaternary center located in the thiazoline portion of the molecule. The structure of bisebromoamide is particularly noteworthy due to its incorporation of six unique subunits, each featuring distinctive chemical characteristics. These include a pivalic acid group, a brominated tyrosine residue, a methylated proline, an α-methyl thiazoline ring, a D-leucine, and a 2-(1-oxo-propyl) pyrrolidine moiety. Bisebromoamide, a novel toxin isolated from an Okinawan strain of *Lyngbya* sp., demonstrates potent anticancer activity with an average IC_{50} value of 40 nM across 39 different cancer cell lines [31]. Notably, its toxicity profile does not align with that of antitubulin agents, as confirmed through biochemical assays. Instead, bisebromoamide exhibits sub-micromolar activity in inhibiting the phosphorylation of extracellular signal-regulated kinase (ERK), highlighting its potential role in disrupting cancer cell signaling pathways [32].

4.5 Largazole

Largazole is a potent natural compound that demonstrates nanomolar-level cytotoxicity against various cancer cell lines, including MDA-MB-231 (breast cancer), NMuMG (mouse mammary gland), and U2OS (osteosarcoma). Its primary molecular target is believed to be histone deacetylases (HDACs), specifically class I HDACs, making it a powerful epigenetic modulator. As an HDAC inhibitor, largazole plays a critical role in regulating gene expression by altering the acetylation status of histone proteins, which can lead to the suppression of cancer cell growth and proliferation [33].

The significance of largazole has prompted the development of several total chemical syntheses, ensuring an ample supply of this compound for further research and biological testing. These synthetic studies have not only confirmed its complex structure but have also facilitated the generation of analogs to explore structure-activity relationships (SAR). This has broadened the understanding of largazole's mechanism of action and enhanced its therapeutic potential. Its dual capabilities as a tool for epigenetic research and a lead compound for anticancer drug development underscore largazole's importance in cancer therapeutics [33].

4.6 Patellamide

Patellamide and a large number of analog structures, collectively known as cyanobactins, have been isolated from marine ascidians which harbor a resident population of the symbiotic cyanobacterium *Prochloron*. These compounds are now understood to be produced by the cyanobacteria through truncation and modification of ribosomally encoded precursor peptides. Cyanobolites represent a highly diverse class of natural products characterized by their potent bioactivities, including anticancer, antimicrobial, and immunosuppressive properties [34]. The biosynthetic machinery involved in cyanobactin production is encoded by conserved gene clusters, which facilitate post-translational modifications such as cyclization, heterocyclization, and oxidative tailoring. This remarkable biosynthetic flexibility enables the generation of structurally unique compounds with significant therapeutic potential. Recent studies have also highlighted the evolutionary adaptations of these symbiotic cyanobacteria, which contribute to their ability to produce such specialized metabolites in marine environments (**Figure 1**) [34].

4.7 Anti-inflammatory activity of cyanobolites different

Several forms of cyanobolites have demonstrated strong and interesting anti-inflammatory effects, with cyanobolites playing a significant role in this recognition. For example, bis-bromoindoles from *Rivularia* sp. have shown anti-inflammatory activity. Recently, marine cyanobacterial metabolites were tested using a nitric oxide (NO) inhibition assay on a mouse RAW macrophage cell line [27]. The results revealed that several malyngamides were powerful inhibitors. Further studies showed

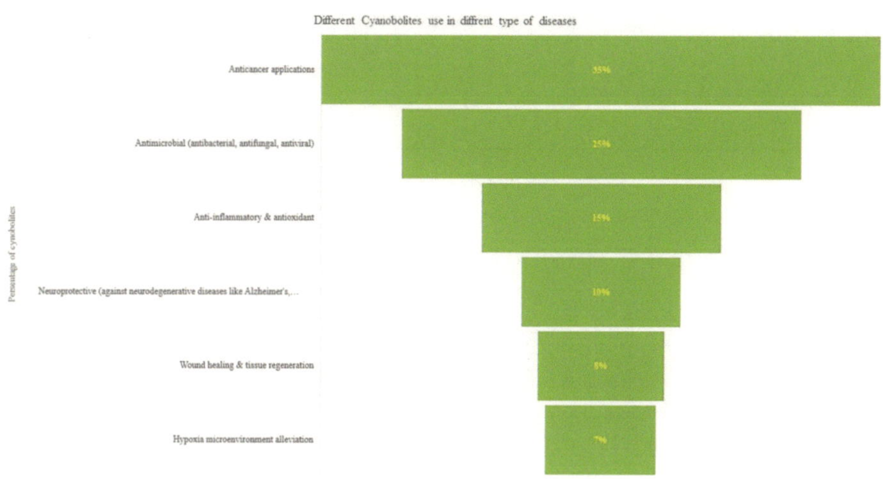

Figure 1.
Cyanobolites percentage use in different disease.

S.N.	Compound name	Species isolate	Mechanism of action	Bioactivity	References
1	Malyngamide 2	*Lyngbya sordida*	Inhibition of nitric oxide (NO) production	Anti-inflammatory activity	[1]
2	Lyngbic Acid	*Lyngbya* species	Linked to cyclohexanone ring *via* an amide group	Anti-inflammatory activity	[2]
3	Oxylipins	*Chlamydomonas debaryana*	Downregulation of cytokines (IL-1β, IL-6, IL-17), inhibition of COX-2, iNOS, NF-κB, increase of HO-1	Anti-inflammatory activity	[3]
4	Astaxanthin	*Haematococcus pluvialis*	Suppression of NF-κB pathway, inhibition of IL-6, IFN-γ expression	Anti-inflammatory activity	[4]
5	Bis-bromoindoles	*Rivularia* sp.	Inhibition of inflammatory pathways	Anti-inflammatory activity	[5]
6	Cylindrospermopsin	*Cylindrospermopsis raciborskii*	Inhibition of protein synthesis	Cytotoxic, hepatotoxic	[6]
7	Microcystin-LR	*Microcystis aeruginosa*	Inhibition of protein phosphatases (PP1 and PP2A)	Hepatotoxic, neurotoxic	[7]
8	Anabaenopeptin	*Anabaena* species	Inhibition of serine proteases	Antiviral, antimicrobial, cytotoxic	[8]
9	Aeruginosin 298-A	*Nostoc* species	Inhibition of protein synthesis	Cytotoxic, antimicrobial	[8]
10	Scytonemin	*Scytonema* species	UV-absorbing, free radical scavenging	Antioxidant, UV protection	[10]
11	BMAA (β-methylamino-L-alanine)	*Aphanizomenon flos-aquae*	Mimics excitatory neurotransmitters and interferes with protein synthesis	Neurotoxic, linked to neurodegenerative diseases	[11]
12	Nostocarboline	*Nostoc* species	Intercalation into DNA, inhibition of DNA replication	Anticancer, antimicrobial	[12]
13	Symplocosin	*Symploca* species	Inhibition of cell division and DNA synthesis	Anticancer	[13]
14	Cyanovirin-N	*Nostoc* species	Binding to HIV gp120, inhibition of viral entry	Antiviral (HIV)	[14]
15	Lyngbya Toxin	*Lyngbya* species	Disruption of cellular membranes	Cytotoxic, neurotoxic	[15]
16	Tetrodotoxin	*Nostoc* species	Blockage of voltage-gated sodium channels	Neurotoxic, paralysis	[16]
17	Oscillatoxin	*Oscillatoria* species	Inhibition of acetylcholinesterase	Anticholinergic, neurotoxic	[17]
18	Aplysiatoxin	*Lyngbya* species	Activation of protein kinase C	Cytotoxic, carcinogenic	[18]
19	Barbamide	*Marine cyanobacteria*	Inhibition of nitric oxide synthase (NOS)	Anti-inflammatory	[20]
20	Dihydroxypropylcysteinylserine	*Nostoc* species	Induction of apoptotic pathways	Anticancer, apoptosis induction	[21]
21	Spumigins	*Nostoc* species	Inhibition of cell proliferation	Anticancer, antimicrobial	[22]

DOI: http://dx.doi.org/10.5772/intechopen.1010149

S.N.	Compound name	Species isolate	Mechanism of action	Bioactivity	References
22	Chlorophyllide A	*Synechocystis* species	Inhibition of photosystem II	Antioxidant, antimicrobial	[23]
23	Muriellin	*Cyanobacterium* sp.	Disruption of cell membranes	Antimicrobial, cytotoxic	[24]
24	Westiellamide	*Westiella* species	Disruption of protein folding and activity	Anticancer, antimicrobial	[25]
25	Kombuamine	*Cyanobacterium* sp.	Inhibition of protein synthesis	Antimicrobial, antiviral	[42]
26	Pyrroloiminoquinone	*Microcystis* species	Inhibition of ribosome function and protein synthesis	Anticancer	[43]
27	Palau'amine	*Marine cyanobacteria*	Inhibition of protein kinase C	Anticancer, neuroprotective	[44]
28	Epinecidin-1	*Marine cyanobacteria*	Induces membrane disruption and cell lysis	Antimicrobial, anticancer	[1]
29	Cyanophycin	*Nostoc* species	Inhibition of bacterial cell wall synthesis	Antimicrobial	[23]
30	Scytovirin	*Scytonema* species	Interference with viral entry (HIV)	Antiviral	[22]
31	Anabaena Alkaloids	*Anabaena* species	Interference with neurotransmitter function	Neurotoxic	[45]
32	Cyanopeptolins	*Microcystis* species	Inhibition of protein phosphatase activity	Cytotoxic, antimicrobial, anticancer	[46]
33	Lyngbyatoxin	*Lyngbya* species	Activation of protein kinase C	Tumor-promoting, carcinogenic	[47]
34	Laurencin	*Marine cyanobacteria*	Inhibition of cell division and protein synthesis	Anticancer	[25]
35	Oscillamide	*Oscillatoria* species	Inhibition of enzyme activity and membrane disruption	Antimicrobial, anticancer	[25]
36	Geosmin	*Anabaena* species	Interference with microbial metabolism	Antimicrobial, affects microbial ecosystem	[48]
37	Hapalosin	*Marine cyanobacteria*	Inhibition of cell division	Anticancer	[1]
38	Dihydroxypropyl cyanotoxin	*Cyanobacterium* sp.	Inhibition of protein synthesis and cell division	Cytotoxic, neurotoxic	[49]
39	Kessylamide	*Kessylama* species	Inhibition of DNA synthesis and repair	Anticancer, antiviral	[50]
40	Ascomycin	*Nostoc* species	Inhibition of calcineurin and immune response	Immunosuppressive	[46]

Table 1.
Different cyanobolites use in different types of biological activity.

that these malyngamides worked by reducing the levels of Interleukin 1 and 6 (IL1 and IL6), while increasing Tumor Necrosis Factor alpha (TNFa). The malyngamides appeared to act by blocking the MyD88 pathway of inflammation, rather than the TRIF pathway. Other malyngamides, like malyngamide 2, also exhibited anti-inflammatory properties [35] (**Table 1**).

4.7.1 Malyngamides

Malyngamides are bioactive compounds primarily identified in the *Lyngbya* genus of cyanobacteria. These compounds typically consist of unsaturated fatty acids, such as Lyngbic acid, which are attached to a cyclohexanone ring *via* an amide group [35]. Among them, Malyngamide 2 is a noteworthy lipopeptide that has been isolated from *Lyngbya sordida*. This compound has demonstrated significant anti-inflammatory effects, particularly in LPS-induced macrophage cells, by inhibiting nitric oxide (NO) production. The anti-inflammatory properties of malyngamides highlight their potential therapeutic applications in treating inflammation-related diseases [36].

4.7.2 Oxylipins

Oxylipins are another group of bioactive compounds found in *Chlamydomonas debaryana*, a species of green microalga. These compounds have shown considerable anti-inflammatory activity, particularly in a colitis model induced by TNBS (2,4,6-trinitrobenzenesulfonic acid) in mice [37]. The biomass of *Chlamydomonas debaryana* was shown to downregulate pro-inflammatory cytokines, including IL-1β, IL-6, and IL-17. Additionally, it decreased the expression of key inflammatory markers such as COX-2, iNOS, and NF-κB while increasing the level of heme oxygenase 1 (HO-1), suggesting a potential mechanism for their anti-inflammatory effects [38].

4.7.3 Astaxanthin

Astaxanthin is a xanthophyll carotenoid found in several organisms, most notably in the microalga *Haematococcus pluvialis*. This compound has gained attention for its potent anti-inflammatory properties. *In vivo* studies have demonstrated that astaxanthin suppresses the NF-κB signaling pathway, leading to reduced expression levels of pro-inflammatory cytokines such as IL-6 and IFN-γ. The ability of astaxanthin to modulate the NF-κB pathway underscores its potential as an effective anti-inflammatory agent for treating inflammatory diseases [35].

4.8 Antimicrobial activity of cyanobolites

The last few decades have seen a rise in multidrug-resistant microbes, leading to a greater emphasis on the discovery and development of new antimicrobial agents with new mechanisms of action. However, recent studies show that only 33% of the developed antibiotics have been successfully translated into commercially available products. Furthermore, the lack of new mechanisms of action and limited novelty in the chemical structure of antibiotics targeting well-established microbe targets have hindered the success of new antibiotic ventures. For this reason, cyanobacteria have attracted attention as a promising source of therapy, producing a diverse array of secondary metabolites with potent antimicrobial activity against various pathogens [39].

4.8.1 Scytonemin

Scytonemin, a unique UV-absorbing compound produced by Scytonema species, not only serves as a protective agent against ultraviolet radiation but also demonstrates antimicrobial properties. By scavenging free radicals and absorbing UV light, Scytonemin reduces microbial proliferation, acting as a natural defense against a range of pathogens. Its dual function as both an antimicrobial agent and a UV protectant makes it an intriguing compound for various applications [36].

4.8.2 Cyanovirin-N

Cyanovirin-N, isolated from Nostoc species, is another fascinating secondary metabolite with potent antiviral activity. This protein binds to the HIV gp120 glycoprotein, blocking viral entry into host cells. Cyanovirin-N has demonstrated broad-spectrum antiviral activity against various viruses and also exhibits antimicrobial properties against bacterial pathogens, making it a promising candidate for developing antiviral and antibacterial therapies [40].

4.8.3 BMAA

BMAA (β-methylamino-L-alanine), produced by Aphanizomenon flos-aquae, is primarily known for its neurotoxic effects, but it also exhibits antimicrobial activity in certain strains. BMAA mimics neurotransmitters and disrupts protein synthesis, which can interfere with bacterial metabolism and growth. While its neurotoxicity is a concern, its antimicrobial properties further highlight the diverse potential of cyanobacterial secondary metabolites [41].

4.8.4 Anabaenopeptin

Another important cyanobacterial metabolite is anabaenopeptin, found in *Anabaena* species. This compound functions by inhibiting serine proteases, essential for bacterial cell functions, thereby exerting antimicrobial activity, especially against gram-positive bacteria. In addition to its antibacterial properties, Anabaenopeptin also displays antiviral effects, making it a versatile antimicrobial agent with significant therapeutic potential [51].

5. Cyanobolites in drug discovery

Cyanobacteria exhibit a remarkable degree of diversity in their bioactivities, largely due to the extensive variety of secondary metabolites they produce [52]. These microorganisms do not specialize in producing a specific class of chemicals; instead, their secondary metabolite profile encompasses a wide range of diverse compounds. This broad spectrum of metabolites can be attributed to their ubiquitous presence across various ecosystems and their long evolutionary history, which has shaped their biochemical versatility (**Figure 2**) [53].

These secondary metabolites display an impressive array of bioactivities, many of which are relevant to their natural environment. Cyanobacterial metabolites have demonstrated antibacterial, antifungal, antiviral, and cytotoxic properties, among others [54]. While some of these functions may be naturally related to the organisms'

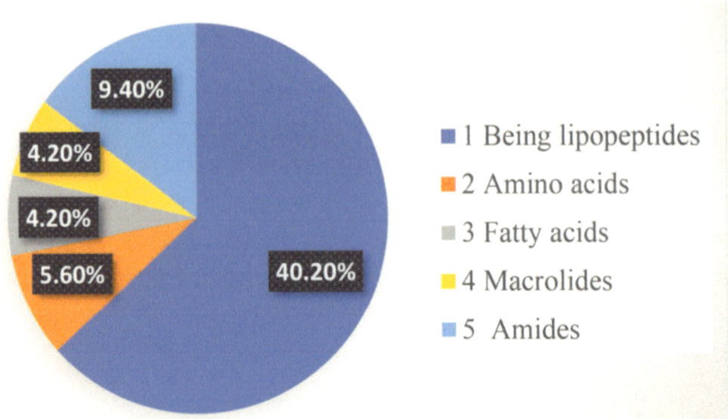

Figure 2.
Cyanobolites present in different cyanobacteria.

ecological niches, others, such as their potential as anticancer agents, immunomodulators, or protease inhibitors, hint at untapped medicinal applications. Thus, cyanobacteria represent a rich source of bioactive compounds with pharmaceutical potential [54].

Recent advancements in genome sequencing have unveiled a remarkable diversity of genes responsible for producing bioactive proteins, including both ribosomal and nonribosomal peptides, as well as peptide–polyketide hybrid molecules. This genetic diversity supports the idea that cyanobacteria are a treasure trove of novel bioactive compounds, many of which could play a significant role in the development of new therapeutic agents. The continued exploration of cyanobacterial genomics and natural products could pave the way for groundbreaking discoveries in pharmaceutical research, particularly in the fields of cancer treatment, immunity modulation, and microbial resistance [55].

6. Molecular mechanisms and *In Silico* approaches

Cyanobacterial secondary metabolites exert their bioactivity through diverse molecular mechanisms, making them valuable for therapeutic applications. These compounds can induce apoptosis by activating caspase cascades, disrupting mitochondrial membrane potential, and modulating Bcl-2 family proteins [56]. They also play a significant role in oxidative stress regulation by either scavenging reactive oxygen species (ROS) or enhancing oxidative damage in target cells, contributing to their cytotoxic effects against cancer and microbial pathogens. Furthermore, these metabolites influence key signal transduction pathways, such as Mitogen-Activated Protein Kinase (MAPK), PI3K/Akt, and NF-κB, which are crucial in cell proliferation, survival, and immune responses. Their ability to inhibit protein kinases, interfere with enzyme activity, and bind to cellular receptors highlights their potential in drug development for combating infectious diseases and cancer [57].

Cyanobacterial secondary metabolites can induce cell cycle arrest by targeting key regulatory proteins and signaling pathways that control cell division. These bioactive compounds interfere with cyclins and cyclin-dependent kinases (CDKs), which are essential for cell cycle progression. For instance, some metabolites inhibit CDK activity, leading to the accumulation of cyclin inhibitors such as p21 and p27, which halt the cell cycle at specific checkpoints like G1/S or G2/M [58]. Additionally, cyanobacterial compounds can activate tumor suppressor proteins like p53, which triggers DNA damage response and promotes cell cycle arrest to prevent the proliferation of damaged or cancerous cells. Some metabolites also disrupt microtubule dynamics, affecting spindle formation during mitosis and leading to mitotic arrest. These mechanisms collectively contribute to the antiproliferative and anticancer properties of cyanobacterial secondary metabolites, making them potential candidates for novel chemotherapeutic agents [59].

Cyanobacterial secondary metabolites modulate Bcl-2 protein activity, playing a crucial role in regulating apoptosis. The Bcl-2 family consists of pro-apoptotic (e.g., Bax and Bak) and anti-apoptotic (e.g., Bcl-2 and Bcl-xL) proteins, which control mitochondrial outer membrane permeabilization (MOMP) and cytochrome c release. Some cyanobacterial compounds act as Bcl-2 inhibitors, promoting apoptosis by disrupting the Bcl-2/Bax interaction, leading to mitochondrial dysfunction and caspase activation [56]. Others enhance the expression of pro-apoptotic proteins like Bax, shifting the balance toward cell death, which is particularly effective in cancer treatment. Additionally, certain metabolites generate oxidative stress, triggering p53 activation and subsequent downregulation of Bcl-2, further sensitizing cells to apoptosis [60]. These mechanisms highlight the therapeutic potential of cyanobacterial metabolites in targeting Bcl-2-mediated pathways for anticancer drug development [56].

Cyanobacterial secondary metabolites play a crucial role in modulating the MAPK (Mitogen-Activated Protein Kinase) and PI3K/Akt (Phosphoinositide 3-Kinase/Akt) signaling pathways, which are essential regulators of cell survival, proliferation, and apoptosis [57, 61]. Many of these bioactive compounds influence the MAPK pathway by either inhibiting or activating specific cascades such as ERK, JNK, and p38 MAPK. Some cyanobacterial metabolites suppress ERK phosphorylation, leading to decreased cancer cell proliferation and enhanced apoptosisn [61]. On the other hand, activation of JNK and p38 MAPK by these compounds promotes pro-apoptotic responses by increasing the expression of Bax, caspases, and cytochrome c release, which drive programmed cell death [62].

Similarly, cyanobacterial metabolites effectively target the PI3K/Akt pathway, a key signaling cascade involved in cancer progression and cell survival. These compounds often function as Akt inhibitors, preventing cancer cells from evading apoptosis. By downregulating phosphorylated Akt (p-Akt), they reduce survival signaling and enhance cellular sensitivity to apoptosis-inducing factors [63]. Additionally, some metabolites suppress mTOR activity, a downstream effector of PI3K/Akt, thereby hindering cancer cell proliferation and promoting autophagy. Moreover, certain cyanobacterial compounds enhance the activity of PTEN, a tumor suppressor that negatively regulates the PI3K/Akt pathway, further reinforcing their anticancer potential. Through these mechanisms, cyanobacterial secondary metabolites emerge as promising candidates for drug development, particularly in cancer therapy, antimicrobial treatments, and anti-inflammatory applications [64].

In silico approaches have transformed the study of cyanobolites by enabling rapid and cost-effective analysis of their biosynthetic potential. Traditional methods of discovering these bioactive compounds, such as culturing and chemical extraction, are often time-consuming and limited by environmental conditions [65]. Computational tools, including genome mining, molecular docking, and metabolic pathway modeling, allow researchers to predict and characterize novel secondary metabolites without the need for extensive laboratory work. Genome sequencing and bioinformatics pipelines help identify biosynthetic gene clusters

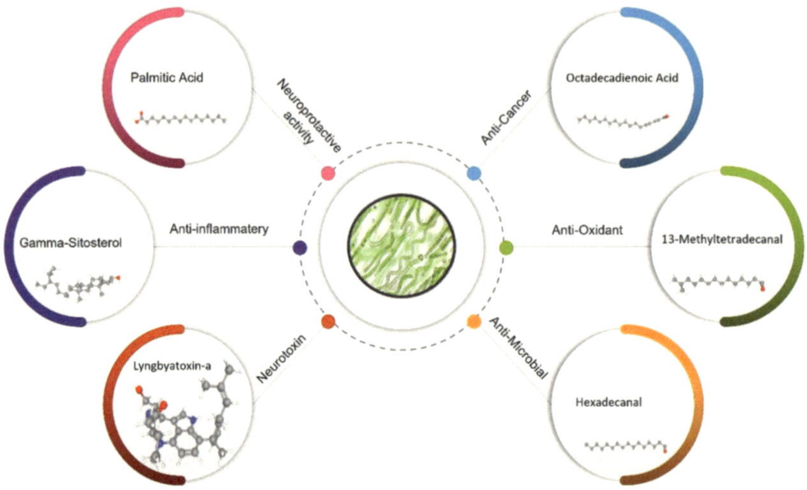

Figure 3.
Cyanobolites use in different types of biological activity.

(BGCs) responsible for producing these compounds, facilitating targeted experimental validation and accelerating the discovery of new natural products with pharmaceutical and industrial applications [66].

Furthermore, *in silico* techniques play a crucial role in understanding the ecological roles and evolutionary patterns of cyanobacterial secondary metabolites. Comparative genomics and machine learning approaches help in deciphering the regulatory mechanisms and genetic diversity of these compounds across different cyanobacterial strains [67]. Molecular docking and virtual screening enable the identification of potential drug candidates by predicting interactions between cyanobacterial metabolites and biological targets. These computational advancements not only enhance drug discovery efforts but also contribute to biotechnological applications, such as bioengineering cyanobacteria for the sustainable production of valuable metabolites. As *in silico* tools continue to evolve, they promise to further revolutionize our ability to harness the vast metabolic potential of cyanobacteria for scientific and commercial benefits (**Figure 3**) [44].

7. Future prospects of cyanobacterial secondary metabolites in biological sciences

Cyanobacterial secondary metabolites hold immense promise for future advancements in biological sciences, particularly in drug discovery, biotechnology, and environmental applications. Their diverse bioactive compounds, including alkaloids, peptides, and polyketides, exhibit potent antimicrobial, antiviral, anticancer, and anti-inflammatory properties, making them valuable candidates for pharmaceutical development. With increasing antibiotic resistance, the exploration of cyanobacterial metabolites for novel antibiotics is gaining momentum. Additionally, their ability to target key molecular pathways like apoptosis, oxidative stress, and signal transduction makes them highly promising for developing next-generation anticancer drugs [68, 69].

The advancement of synthetic biology has significantly expanded the biomedical potential of cyanobacteria, particularly through the use of living or engineered cyanobacteria. This offers promising strategies for future disease treatments. To better understand and enable these applications, we have highlighted the key properties of cyanobacteria relevant to biomedical use, such as their production of bioactive compounds and their ability to adsorb heavy metals. Building on these properties, we explored their growing role in various disease models, including the alleviation of hypoxic microenvironments, wound healing, and drug delivery. Looking ahead, we discussed future prospects, including further exploration of cyanobacterial secondary metabolites, integrating bioactive compounds synthesized by cyanobacteria directly into medical diagnostics and treatments, and optimizing their *in vivo* applications [43].

Beyond medicinal drugs, cyanobacterial secondary metabolites are being explored for their potential in bioremediation and sustainable agriculture. Certain metabolites show allelopathic effects, inhibiting the growth of harmful microbes and pests, which can be harnessed for eco-friendly biofertilizers and biopesticides. Furthermore, their role in nanotechnology is emerging, where cyanobacteria-derived compounds are used in green synthesis of nanoparticles for medical and industrial applications. Advances in genetic engineering and synthetic biology may also enable the enhancement of metabolite production, making large-scale utilization feasible. As research continues to unveil the vast biochemical diversity of cyanobacteria, the integration of genomic, metabolomic, and biotechnological approaches will be crucial in fully harnessing their potential. Future studies focusing on mechanistic insights, bioavailability, and clinical applications will pave the way for transforming cyanobacterial metabolites into mainstream therapeutic and industrial solutions, positioning them as vital resources in the biological sciences [43].

8. Challenges and opportunities

8.1 Challenges of cyanobolites

Cyanobolites, derived from cyanobacteria, present several challenges that limit their widespread application and research potential. One major challenge is the complexity and variability of cyanobacterial metabolites, which can vary significantly depending on environmental conditions such as light, temperature, and nutrient availability. This variability complicates the standardization and reproducibility of bioactive compounds for pharmaceutical or industrial applications. Additionally, cultivating cyanobacteria on a large scale can be difficult due to the need for specific growth conditions and the risk of contamination by other microorganisms. The extraction and purification of cyanobolites also require advanced techniques, adding to the cost and technical challenges. Moreover, regulatory hurdles in approving new bioactive compounds from cyanobacteria for therapeutic use further impede their integration into mainstream applications [68].

8.2 Opportunities of cyanobolites

Despite the challenges, cyanobolites offer substantial opportunities in various fields due to their diverse biological activities, including anticancer, antimicrobial, and antioxidant properties. Advances in biotechnological methods, such as genetic engineering and metabolic pathway optimization, have the potential to enhance the production of valuable cyanobolites, making them more accessible for commercial use [65]. The increasing interest in natural and sustainable bioactive compounds drives research into cyanobolites, presenting opportunities for developing novel pharmaceuticals and eco-friendly products. Additionally, the unique chemical structures of cyanobolites open avenues for the discovery of new drugs with mechanisms distinct from existing therapies, addressing antibiotic resistance and providing new treatment options for various diseases. The integration of *in silico* analysis and high-throughput screening further accelerates the identification and development of promising cyanobolites, expanding their potential applications [44].

9. Conclusion

In conclusion, the ability of cyanobacteria to produce a wide range of bioactive secondary metabolites, known as cyanobolites, is a key factor in their survival and ecological success. These compounds, including peptides, alkaloids, polyketides, and terpenoids, play important roles in stress responses, such as protection from predators, UV radiation, and oxidative damage. Their diverse biological activities—ranging from antimicrobial to anticancer properties—highlight their potential for biotechnological applications in drug discovery and environmental remediation. In particular, the interactions between nonribosomal peptides (NRPs) and polyketides observed in compounds such as microcystins underscore their importance in the development of new bioactive agents. Despite significant progress, many cyanobacterial metabolites are still underexplored, presenting vast opportunities for future research. Advances in molecular mechanisms and *in silico* approaches will be instrumental in uncovering new compounds and their therapeutic potentials, paving the way for new developments in biological and medical sciences.

Moreover, with further research, cyanobolites could become an important resource to address global challenges such as antimicrobial resistance, cancer therapy, and environmental pollution. Their potential for use in agricultural biotechnology—as natural pesticides and growth promoters—also opens up new sustainable solutions for food production. As the demand for environmentally friendly, cost-effective bioproducts increases, the discovery of cyanobacterial metabolites for industrial applications becomes more important. The continued discovery of these metabolites offers promising avenues for innovation in pharmaceuticals, agriculture, and sustainability. In the long term, these bioactive compounds could contribute to personalized medicine, environmental protection, and the development of new biotechnological processes that could benefit both human health and mother earth.

Acknowledgements

The authors would like to express their sincere gratitude to Guru Ghasidas Vishwavidyalaya, Bilaspur, for providing an intellectually stimulating environment and the necessary resources to carry out this research. The support from the university has been invaluable in facilitating the completion of this study.

Conflict of interest

The Author has declared no conflict of interest.

Author details

Kundan Ojha[1], Satish Dubey[1], Akanksha Singh[1], Ashwini Kumar Dixit[2]*
and Priyanka Kumari[1]

1 Laboratory of Molecular Taxonomy and Medicinal Plant Biology, Department of Botany, Guru Ghasidas Vishwavidyalaya, Bilaspur, Chhattisgarh, India

2 Department of Botany, Guru Ghasidas Vishwavidyalaya, Bilaspur, Chhattisgarh, India

*Address all correspondence to: dixitak@live.com

References

[1] Sreeharsha RV, Venkata MS. Bacterial photosynthesis for nutrient recovery and value addition. In: Microbial Photosynthesis: From Basic Biology to Artificial Cell Factories and Industrial Applications. Singapore: Springer Nature Singapore; 2024. pp. 125-136

[2] Harris TD, Reinl KL, Azarderakhsh M, Berger SA, Berman MC, Bizic M, et al. What makes a cyanobacterial bloom disappear? A review of the abiotic and biotic cyanobacterial bloom loss factors. Harmful Algae. 2024;**131**:102599

[3] Liu D, Liberton M, Hendry JI, Aminian-Dehkordi J, Maranas CD, Pakrasi HB. Engineering biology approaches for food and nutrient production by cyanobacteria. Current Opinion in Biotechnology. 2021;**67**:1-6

[4] Narayan KP, Tiwari S, Pabbi S, Dhar DW. Biodiversity analysis of selected cyanobacteria. Current Science. 2006;**91**(7):1-8

[5] Nayak S, Prasanna R, Prasanna BM, Sahoo DB. Analysing diversity among Indian isolates of anabaena (Nostocales, Cyanophyta) using morphological, physiological and biochemical characters. World Journal of Microbiology and Biotechnology. 2007;**23**:1575-1584

[6] Cotas J, Lomartire S, Gonçalves AM, Pereira L. From ocean to medicine: Harnessing Seaweed's potential for drug development. International Journal of Molecular Sciences. 2024;**25**(2):797

[7] Burja AM et al. Marine cyanobacteria—a prolific source of natural products. Tetrahedron. 2001;**57**(46):9347-9377

[8] Gupta V, Ratha SK, Sood A, Chaudhary V, Prasanna R. New insights into the biodiversity and applications of cyanobacteria (blue-green algae)—Prospects and challenges. Algal Research. 2013;**2**(2):79-97

[9] Kehr JC, Picchi DG, Dittmann E. Natural product biosyntheses in cyanobacteria: A treasure trove of unique enzymes. Beilstein Journal of Organic Chemistry. 2011;**7**(1):1622-1635

[10] Dixit RB, Suseela MR. Cyanobacteria: Potential candidates for drug discovery. Antonie Van Leeuwenhoek. 2013;**103**:947-961

[11] Kaur G. Freshwater cyanotoxins. In: Biomarkers in Toxicology. Academic Press; 2019. pp. 601-613

[12] Sivonen K, Börner T. Bioactive compounds produced by cyanobacteria. Applied Microbiology and Biotechnology. 2010

[13] Engene N, Choi H, Esquenazi E, Rottacker EC, Ellisman MH, Dorrestein PC, et al. Underestimated biodiversity as a major explanation for the perceived rich secondary metabolite capacity of the cyanobacterial genus Lyngbya. Environmental Microbiology. 2011;**13**(6):1601-1610

[14] Nowruzi B, Porzani SJ. Toxic compounds produced by cyanobacteria belonging to several species of the order Nostocales: A review. Journal of Applied Toxicology. 2021;**41**(4):510-548

[15] Cock IE, Cheesman MJ. A review of the antimicrobial properties of cyanobacterial natural products. Molecules. 2023;**28**(20):7127

[16] Singh S, Chib S, Akhtar MJ, Kumar B, Chawla PA, Bhatia R.

Paradigms and success stories of natural products in drug discovery against neurodegenerative disorders (NDDs). Current Neuropharmacology. 2024;**22**(6):992-1015

[17] Bhatla SC, Lal MA. Secondary metabolites. In: Physiology P, editor. Development and Metabolism. Singapore: Springer Nature Singapore; 2023. pp. 765-808

[18] Khan F, Akhlaq A, Rasool MH, Srinuanpan S. Cyanobacterial bioactive compounds: Synthesis, extraction, and applications. In: Pharmaceutical and Nutraceutical Potential of Cyanobacteria. Cham: Springer International Publishing; 2024. pp. 215-243

[19] Nunnery JK, Mevers E, Gerwick WH. Biologically active secondary metabolites from marine cyanobacteria. Current Opinion in Biotechnology. 2010;**21**(6):787-793

[20] Bouyahya A, Bakrim S, Chamkhi I, Taha D, El Omari N, El Mneyiy N, et al. Bioactive substances of cyanobacteria and microalgae: Sources, metabolism, and anticancer mechanism insights. Biomedicine and Pharmacotherapy. 2024;**170**:115989

[21] Irankhahi P, Riahi H, Batool Hassani S, Eskafi M, Irani MA, Shariatmadari Z. Formation of the UV-C Protection Shield in Cyanobacterium Nostoc; Investigating the Photoprotective Compounds, Expression of Biosynthesis Genes and Molecular Interactions. Sharjah, United Arab Emirates: Bentham Science Publishers B.V; 2024

[22] Nawaz T, Saud S, Gu L, Khan I, Fahad S, Zhou R. Cyanobacteria: Harnessing the power of microorganisms for plant growth promotion, stress alleviation, and phytoremediation in the era of sustainable agriculture. Plant. Stress. 2024:100399

[23] Rezayian M, Niknam V, Ebrahimzadeh H. Stress response in cyanobacteria. Iranian Journal of Plant Physiology. 2019;**9**(3):2773-2787

[24] Rigonato J, Alvarenga DO, Fiore MF. Tropical cyanobacteria and their biotechnological applications. Diversity and Benefits of Microorganisms from the Tropics. 2017:139-167

[25] Hwang S, Lee N, Cho S, Palsson B, Cho BK. Repurposing modular polyketide synthases and non-ribosomal peptide synthetases for novel chemical biosynthesis. Frontiers in Molecular Biosciences. 2020;**7**:87

[26] Nogle LM, Gerwick WH. Somocystinamide A, a novel cytotoxic disulfide dimer from a Fijian marine cyanobacterial mixed assemblage. Organic Letters. 2002;**4**(7):1095-1098

[27] Nowruzi B, Sarvari G, Blanco S. The cosmetic application of cyanobacterial secondary metabolites. Algal Research. 2020;**49**:101959

[28] Hussein HA, Khaphi FL, Saeed ZK. Cytotoxicity of bioactive compounds derived from cyanobacteria. INNOSC Theranostics and Pharmacological Sciences. 2023;**X**(X):1-12

[29] Skiba MA, Sikkema AP, Moss NA, Lowell AN, Su M, Sturgis RM, et al. Biosynthesis of t-butyl in apratoxin A: Functional analysis and architecture of a PKS loading module. ACS Chemical Biology. 2018;**13**(6):1640-1650

[30] Shi W, Lu D, Wu C, Li M, Ding Z, Li Y, et al. Coibamide A kills cancer cells through inhibiting autophagy. Biochemical and Biophysical Research Communications. 2021;**547**:52-58

[31] Suzuki K, Mizuno R, Suenaga K, Teruya T, Tanaka N, Kosaka T, et al. B

isebromoamide, an extract from Lyngbya species, induces apoptosis through ERK and mTOR inhibitions in renal cancer cells. Cancer Medicine. 2013;**2**(1):32-39

[32] Robles-Bañuelos B, Durán-Riveroll LM, Rangel-López E, Pérez-López HI, González-Maya L. Marine cyanobacteria as sources of lead anticancer compounds: A review of families of metabolites with cytotoxic, antiproliferative, and antineoplastic effects. Molecules. 2022;**27**(15):4814

[33] Hong J, Luesch H. Largazole: From discovery to broad-spectrum therapy. Natural Product Reports. 2012;**29**(4):449-456

[34] Demay J, Bernard C, Reinhardt A, Marie B. Natural products from cyanobacteria: Focus on beneficial activities. Marine Drugs. 2019;**17**(6):320

[35] Pagels F, Guedes AC, Vasconcelos V, Lopes G. Anti-inflammatory compounds from cyanobacteria. In: The Pharmacological Potential of Cyanobacteria. Frankfurt am Main, Germany: Beilstein-Institut zur Förderung der Chemischen Wissenschaften; 2022. pp. 81-105

[36] Srivastava R, Prajapati R, Kanda T, Yadav S, Singh N, Yadav S, et al. Phycochemistry and bioactivity of cyanobacterial secondary metabolites. Molecular Biology Reports. 2022;**49**(11):11149-11167

[37] de Los RC, Ávila-Román J, Ortega MJ, de la Jara A, García-Mauriño S, Motilva V, et al. Oxylipins from the microalgae Chlamydomonas debaryana and Nannochloropsis gaditana and their activity as TNF-α inhibitors. Phytochemistry. 2014;**102**:152-161

[38] Tabarzad M, Atabaki V, Hosseinabadi T. Anti-inflammatory activity of bioactive compounds from microalgae and cyanobacteria by focusing on the mechanisms of action. Molecular Biology Reports. 2020;**47**(8):6193-6205

[39] Tyagi R, Kaushik BD, Kumar J. Antimicrobial activity of some cyanobacteria. Microbial Diversity and Biotechnology Food Security. 2014:463-470

[40] Grabowski Ł, Wiśniewska K, Żabińska M, Konarzewska M, Zielenkiewicz M, Rintz E, et al. Cyanobacteria and their metabolites-can they be helpful in the fight against pathogenic microbes? Blue Biotechnology. 2024;**1**(1):4

[41] de Freitas PNN, Jacinavicius FR, Passos LS, de Souza AO, Dextro RB, Pinto E. Exploring the biodiversity of Antarctic cyanobacteria: A review of secondary metabolites and their applications. Algal Research. 2024;**103617**

[42] Moore MN. Do airborne biogenic chemicals interact with the PI3K/Akt/mTOR cell signalling pathway to benefit human health and wellbeing in rural and coastal environments? Environmental Research. 2015;**140**:65-75

[43] Zhang J, Zhang F, Dong Z, Zhang W, Sun T, Chen L. Response and acclimation of cyanobacteria to acidification: A comprehensive review. Science of the Total Environment. 2024;**173978**

[44] Sahu N, Mishra S, Kesheri M, Kanchan S, Sinha RP. Identification of cyanobacteria-based natural inhibitors against SARS-CoV-2 druggable target ACE2 using molecular docking study, ADME and toxicity analysis. Indian Journal of Clinical Biochemistry. 2023;**38**(3):361-373

[45] Mandhata CP, Bishoyi AK, Sahoo CR, Maharana S, Padhy RN. Insight to biotechnological utility of phycochemicals from cyanobacterium anabaena sp.: An overview. Fitoterapia. 2023;**169**:105594

[46] Köcher S, Resch S, Kessenbrock T, Schrapp L, Ehrmann M, Kaiser M. From dolastatin 13 to cyanopeptolins, micropeptins, and lyngbyastatins: The chemical biology of Ahp-cyclodepsipeptides. Natural Product Reports. 2020;**37**(2):163-174

[47] Edwards DJ, Gerwick WH. Lyngbyatoxin biosynthesis: Sequence of biosynthetic gene cluster and identification of a novel aromatic prenyltransferase. Journal of the American Chemical Society. 2004;**126**(37):11432-11433

[48] Micallef ML, D'Agostino PM, Sharma D, Viswanathan R, Moffitt MC. Genome mining for natural product biosynthetic gene clusters in the subsection V cyanobacteria. BMC Genomics. 2015;**16**:1-20

[49] El Amrani ZS, El Khalloufi F, Mugani R, El Mahdi R, Kasrati A, Soulaimani B, et al. Seaweed essential oils as a new source of bioactive compounds for cyanobacteria growth control: Innovative ecological biocontrol approach. Toxins. 2020;**12**(8):527

[50] Méjean A, Ploux O. A genomic view of secondary metabolite production in cyanobacteria. Advances in Botanical Research. 2013;**65**:189-234

[51] Vestola J, Shishido TK, Jokela J, Fewer DP, Aitio O, Permi P, et al. Hassallidins, antifungal glycolipopeptides, are widespread among cyanobacteria and are the end-product of a nonribosomal pathway. National Academy of Sciences of the United States of America. 2014;**111**(18):E1909-E1917

[52] Rastogi RP, Sinha RP. Biotechnological and industrial significance of cyanobacterial secondary metabolites. Biotechnology Advances. 2009;**27**(4):521-539

[53] Gademann K, Portmann C. Secondary metabolites from cyanobacteria: Complex structures and powerful bioactivities. Current Organic Chemistry. 2008;**12**(4):326-341

[54] Shah SAA, Akhter N, Auckloo BN, Khan I, Lu Y, Wang K, et al. Structural diversity, biological properties and applications of natural products from cyanobacteria: A review. Marine Drugs. 2017;**15**(11):354

[55] Nikolouli K, Mossialos D. Bioactive compounds synthesized by non-ribosomal peptide synthetases and type-I polyketide synthases discovered through genome-mining and metagenomics. Biotechnology Letters. 2012;**34**:1393-1403

[56] Gupta A, Singh PR, Singh AP, Kumari N, Jaiswal J, Sahu N, et al. Anticancer compounds from cyanobacteria and their implications in apoptosis. Current Protein and Peptide Science. 2023;**24**(10):805-819

[57] Huang JJ, Xu W, Lin S, Cheung PCK. The bioactivities and biotechnological production approaches of carotenoids derived from microalgae and cyanobacteria. Critical Reviews in Biotechnology. 2024;**1-29**

[58] Freitas S, Martins R, Campos A, Azevedo J, Osório H, Costa M, et al. Insights into the potential of picoplanktonic marine cyanobacteria strains for cancer therapies–cytotoxic mechanisms against the RKO colon cancer cell line. Toxicon. 2016;**119**:140-151

[59] Štern A, Rotter A, Novak M, Filipič M, Žegura B. Genotoxic effects

of the cyanobacterial pentapeptide nodularin in HepG2 cells. Food and Chemical Toxicology. 2019;**124**:349-358

[60] Bhattacharjee S, Singh P, Bhardwaj A, Mishra AK. Cell death in photoautotrophs. In: Stress Biology in Photosynthetic Organisms: Molecular Insights and Cellular Responses. Singapore: Springer Nature Singapore; 2024. pp. 385-410

[61] Liao Y, Deng Y, Yu X, Zhang P, Liu R. The mediating role of AKT/ERK/JNK signaling on the malignant phenotype of microcystin-LR in gastric adenocarcinoma cells. Food and Chemical Toxicology. 2023;**182**:114174

[62] Meng G, Sun Y, Fu W, Guo Z, Xu L. Microcystin-LR induces cytoskeleton system reorganization through hyperphosphorylation of tau and HSP27 via PP2A inhibition and subsequent activation of the p38 MAPK signaling pathway in neuroendocrine (PC12) cells. Toxicology. 2011;**290**(2-3):218-229

[63] Xie H, Kjellström J, Lindblad P. Sustainable production of photosynthetic isobutanol and 3-methyl-1-butanol in the cyanobacterium Synechocystis sp. PCC 6803. Biotechnology for Biofuels and Bioproducts. 2023;**16**(1):134

[64] Joshi V, Bharadvaja N. Current prospects and clinical status of microalgae derived chemotherapeutics. Revista Brasileira de Farmacognosia. 2023;**33**(3):445-470

[65] Wase NV, Wright PC. Systems biology of cyanobacterial secondary metabolite production and its role in drug discovery. Expert Opinion on Drug Discovery. 2008;**3**(8):903-929

[66] Alam K, Islam MM, Gong K, Abbasi MN, Li R, Zhang Y, et al. In silico genome mining of potential novel biosynthetic gene clusters for drug discovery from Burkholderia bacteria. Computers in Biology and Medicine. 2022;**140**:105046

[67] Jones MR, Pinto E, Torres MA, Dörr F, Mazur-Marzec H, Szubert K, et al. CyanoMetDB, a comprehensive public database of secondary metabolites from cyanobacteria. Water Research. 2021;**196**:117017

[68] Singh R, Parihar P, Singh M, Bajguz A, Kumar J, Singh S, et al. Uncovering potential applications of cyanobacteria and algal metabolites in biology, agriculture and medicine: Current status and future prospects. Frontiers in Microbiology. 2017;**8**:515

[69] Verma S, Thapa S, Siddiqui N, Chakdar H. Cyanobacterial secondary metabolites towards improved commercial significance through multiomics approaches. World Journal of Microbiology and Biotechnology. 2022;**38**(6):100

Chapter 5

Flavonoids: A Natural Shield of Plants under Drought Stress

Jajati Keshari Nayak, Dipsikha Mohanty, Debashis Mahapatra, Subhankar Mondal, Ashish Mohanty, Sushree Sangeeta, Gyanisha Nayak, Jeetendra Senapaty and Babyrani Panda

Abstract

Drought stress is a critical environmental stress and substantially restricts crop growth by lowering plant biomass, quality, and energy. It is a dangerous ecological limit induced by the changing temperatures, unevenness of light radiation, and unpredictable rain. Flavonoids as a large class of polyphenolic secondary metabolites, are associated with alleviating damaging effects of drought stress. These are dispersed in various plant parts like leaves, flowers, seeds, and bark. These compounds act as potent antioxidants, which scavenge ROS generated upon different stressful conditions and subsequently reduce the oxidative damage to cellular components. Flavonoids also donate to enhanced salinity stress tolerance through crucial physiological mechanisms, including stomatal closure, osmotic adjustment, as well as transduction pathways connected with the stress responses. However, little attention has been paid to the study of flavonoids as a qualitative response to drought stress.

Keywords: antioxidant, drought stress, flavonoids, plant defense, reactive oxygen species

1. Introduction

Abiotic factors being a primitive component of the environment determine the growth and development of plants [1]. In nature, plants are highly susceptible to various challenging environmental conditions likely cold, heat, drought, flood, salinity, heavy metal stresses, etc. which substantially reduce the productivity of plants [2]. Among the abiotic stresses, drought stress is a persistent and significant issue that hinders plant growth and development accounting for about 70% of potential loss in crop yield [3]. The repercussions of drought stress on the morphology of plants can be visible throughout the growth stage. Yield loss of up to 40% has been recorded in many crops due to drought stress such as 40% in wheat and 21% in maize [4] and 34–68% in cowpea [5].

Drought stress imposes reduced leaf water potential, turgor pressure, and stomatal functioning which impair the expansion and proliferation of cells. It disrupts the photosynthetic machinery, reduces respiration, ion uptake, and translocation, and increases oxidative stress in plants. Due to this, stunted plant growth can be seen, with increased leaf rolling, increased root-to-shoot ratio, reduced plant biomass, longevity, and early maturity causing yield losses [6]. Plants under drought stress typically develop three mechanisms: tolerance, avoidance, and escape. In drought escape plants use developmental flexibility, whereas higher water absorption and decreased water loss are associated with drought avoidance. In the tolerance mechanism, plants exhibit enhanced Osmo protection, antioxidant activity, and greater dehydration tolerance [7].

In drought stress, oxidative stress is considered to be fatal due to the production of various harmful radicals such as superoxide anion radicals (O_2^-), hydrogen peroxide (H_2O_2), singlet oxygen ($^1O_2$), and hydroxyl radicals (OH) [8]. These ROS are highly reactive molecules that oxidize nucleic acid, protein, and carbohydrates. Chloroplast, mitochondria, and peroxisomes are the potential sites of ROS production in plants as a by-product of aerobic metabolism. In drought stress excess production of ROS mainly causes lipid peroxidation and damage to the photosynthetic system [9, 10]. To deal with the oxidative stress damage plants activate a range of antioxidant mechanisms that nullify the detrimental effect of ROS. This antioxidant system includes enzymatic antioxidants like glutathione reductase, glutathione peroxidase, ascorbate peroxidase, superoxide dismutase, catalase, etc. as well as non-enzymatic antioxidants like α-tocopherol, glutathione, carotenoid, and polyphenols including phenolic acid and flavonoids [11]. The role of the enzymatic antioxidant system in drought mitigation has been extensively studied before [12, 13]. However, the drought tolerance mechanism by non-enzymatic antioxidants like flavonoids is still elusive.

Several shreds of evidence suggest that drought stress can stimulate the synthesis of bioactive metabolites such as flavonoids [1, 14]. Flavonoids are a diverse group of secondary metabolites with multifaceted roles in plants. More than 10,000 structures have been identified to date and are well distributed in natural resources like fruits, vegetables, flowers, roots, stems, bark, leaves, etc. [15]. These are low molecular weight polyphenolic compounds and are known as specialized metabolites [16]. It is worth highlighting that, the structural diversity of flavonoids allows them to interact with a wide array of biomolecules [17]. Flavonoids are associated with a wide range of functions in plants including flower color, insect attraction, pollination, UV protection, signaling, male fertility, allelopathy, auxin transport, etc. [15]. Flavonoid biosynthetic genes are found to be increased under various biotic and abiotic stress conditions, especially drought. The main function of flavonoids in drought stress tolerance is to prevent oxidative damage and maintain homeostasis in plants [18]. Flavonoids constitute an efficient non-enzymatic antioxidant system. Under severe drought stress when the enzymatic antioxidant system fails to cope with the degree of oxidation and ROS generation plant activates the non-enzymatic antioxidant system involving flavonoids to counterbalance the excess ROS generation [16, 19]. Flavonoids are mostly found in the mitochondria, chloroplast, and vacuoles which are the potential site of ROS generation. This suggests the putative role of flavonoids as ROS quenchers [20].

The involvement of flavonoids in mitigating oxidative damage in drought stress conditions has been reported before. However, the precise mechanism by which flavonoids contribute to drought stress tolerance still requires further study. In this chapter, we will discuss the exact mechanism behind flavonoid-mediated drought stress tolerance at the physiological and molecular levels.

2. Classifications of flavonoids

Plants secret secondary metabolites for growth, development, and defense mechanisms. It is accumulated in vacuoles in the form of glycosides. According to its basic structure, it has three rings C6-C3-C6. Based on its structure, glycosylation, and acylation, it is classified into seven groups: flavanols, flavones, isoflavones, anthocyanidins, flavanones, flavanols, and chalcones.

2.1 Flavanols

Flavanols (flavan-3-ols) are monomers. They are distinguished by the binding of the OH^- group to the 3rd Carbon of the skeleton. The double bond is absent between the 2nd and 3rd carbon of the ring and the OH group is variable in different flavanols. Generally, flavanols are abundant in fruit skins (e.g., apples, plums, berries, grapes) which accelerates the endothelial system and prohibits the occurrences of cardiovascular diseases [21, 22].

2.2 Anthocyanins

Naturally, anthocyanidins are unstable whereas their glycosylated form (anthocyanins) are stable in plants [23]. Chemically, these are the derivatives of polymethoxy and polyhydroxy. In plants, more than 650 compounds are identified among them most abundant are delphinidin, cyanidin, petunidin, peonidin, malvidin, and pelargonidin [24]. Mainly, these compounds are found in fruits and vegetables (tomato, potato, eggplant, blueberries, carrot). The acylation of anthocyanins has higher antioxidant activities than the glycosylated form. The natural blue colorant is generally derived from the anthocyanins of butterfly peas [25]. Mono-acylated anthocyanins derive from carrots and these are used in the food colorant industry [26]. Anthocyanins are well known for their antioxidant activities. Different anthocyanins compound has different strategies for scavenging nitrogen or ROS species [27]. Anthocyanidins have higher antioxidant activities than anthocyanins due to their structural features [28]. The intensity of antioxidant activities depends upon ring orientation, free OH groups around the pyrone ring, and their positions through orthohydroxylation and methoxylation.

2.3 Flavanones

Abundant in Rutaceae family like citrus fruits (orange and lemon). These are well known as dihydroflavanones. The double bond is present between C_2—C_3 which is a differentiating character of flavanones. In citrus, aglycone is found which can be combined with various glucosides to obtain different flavanones. The most representative flavanones are eriodictyol in lemons, naringenin in grapefruit, and hesperidin and naringin in oranges. In general, citrus fruit varieties vary in their flavanone content [29]. Flavonoids and anthocyanins are both prevalent in red citrus varieties. Flavanones can produce phenolic hydrogen, which makes them an efficient antioxidant. Catalase (CAT), glutathione peroxidase (GPx), paraoxonase (PON), superoxide dismutase (SOD), and other different antioxidant enzymes can all be activated by naringin [30].

2.4 Flavonols

Flavonols (3-hydroxyflavone) are characterized by the A and B rings substituted by various groups and ultimately joined by a 3C chain. It has the OH group in the 5th and 7th positions of the chain. It has more 3-OH group than other flavonoids. These pigments are found more in plant epidermal cells and protect from UV rays [31]. Further, these compounds are classified as galangin, myricetin, quercetin, and kaempferol. Vegetables like onions, broccoli, and fruits like apples are rich in flavonols.

2.5 Isoflavones

These polyphenolic compounds are based on the backbone of 3-phenyl chromene-4-one [32]. Primarily found in the leguminous crops. The primary sources of isoflavones are soy-based products, chickpeas, and alfalfa. There are 12 isomers of soybean isoflavones, including genistein and daidzein [33]. Based on their structural resemblance to estradiol-17β and molecular resemblance to animal estrogens, isoflavones also serve as phytoestrogens [34].

2.6 Flavones

Flavones represent one of the most prominent groups of flavonoids. A phenyl substituent is present at position 2 of 4H-chromen-4-one, which makes up the flavone chemical structure. Most flavones are 7-O-glycosides, which can be found in oranges, red pepper, celery, and tea. Lutein and apigenin are the two primary edible flavones. The tricyclic core structure of apigenin is usually glycosylated, with the sugar moiety being connected to it either directly (C-glycosides) or through hydroxyl groups (O-glycosides). The members of the apigenin group are rhoifolin, schaftoside, vitexin, isovitexin, apiin, and apigenin-7-O-glucoside [35].

2.7 Chalcones

These are the naturally occurring open-chain flavonoids conjugated with prenyl moieties in the C5, C10, and C15 positions of the A and B rings. These bioactive compounds are more prominent in Zingiberaceae, Moraceae, Cannabaceae, and Fabaceae families [14]. xanthohumol and isobavachalcone are the most abundant derivates of chalcone [36]. Flavonoids and isoflavonoids are also derived from it. From basic aromatic compounds, their structural features can be easily constructed. The synthesis of chalcone analogs and slight structural alterations of natural chalcones have been motivated by their notable bioactivity [37].

3. Flavonoid biosynthesis

Flavonol synthase (FLS) is a pivotal enzyme that produces various flavonols such as myricetin, quercetin, and kaempferol from dihydroflavonol. Furthermore, this dihydroflavonol is also reduced by dihydroflavonol 4-reductase (DFR) enzyme to produce leucoanthocyanidins, which are subsequently converted into anthocyanidins by leucoanthocyanidin dioxygenase (LDOX). Further, uridine diphosphate (UDP)-glucose flavonoid-3-O-glycosyltransferase (UFGT) catalyzes to produce the glycosylated form of anthocyanidins. Leucoanthocyanidins can also be converted into flavanols through the action of leucoanthocyanidin reductase (LAR), leading to the formation of proanthocyanidin oligomers or polymers [38]. Finally, processes like acylation, methylation, and glycosylation modify various anthocyanins and flavonols, resulting in stable flavonoid compounds that accumulate in plants [39].

Flavonoid biosynthesis is a combined process including the shikimate pathway (for coumarin CoA synthesis substrate) and the acetate pathway (for malonyl-CoA). Both pathways provide essential components to chalcone synthase for the synthesis of chalcones (the first product in the flavonoid biosynthetic pathway). Chalcone serves as a precursor for the synthesis of other types of flavonoids like flavones, anthocyanins, and flavonols [40]. The first product of the shikimate pathway is phenylalanine, which forms cinnamic acid via phenylalanine ammonia-lyase (PAL) catalysis. At a subsequent time, p-coumaric acid is formed from hydroxylation of the fourth Carbon position of cinnamic acid through cinnamic acid hydroxylase [41].

Upon condensation with the ligase enzyme, Coumarin acid is converted into p-coumaroyl CoA. Further, through isomerization, p-coumaroyl CoA and Malonyl CoA produce naringenin chalcone [42]. Naringenin is an imperative precursor of most flavonoid synthesis and operates the structural similarity among various types of flavonoids. Naringenin can catalyze the formation of flavones by reacting with flavone synthase I (FNS I) or II (FNS II), or it can form isoflavones by reacting with isoflavone synthase (IFS). Additionally, flavanone-3-hydroxylase (F3H) can convert naringenin to dihydrokaempferol. Dihydroquercetin and dihydromyricetin are produced by chemical reactions with flavonol 3′ -hydroxylase and flavonol 3′ 5′ -hydroxylase respectively. Lastly, acylation, methylation, and glycosylation can be used to modify different kinds of anthocyanins and flavonols to create relatively stable flavonoids that can build up in plants [43]. The flavonoid synthesis is a complex process. A series of enzymes like chalcone synthase, chalcone isomerase, and flavonol synthase control this metabolic pathway. Some structural genes control flavonoid synthesis have been identified in apples, rice, and other plants [44].

3.1 Transcription factor (TF) mediated biosynthesis

The plant TF are classified based upon their DNA-binding domains into families like MYB, WRKY, bHLH, NAC, and bZIP. These families play a pivotal role during plant growth, development, and biotic and abiotic stress management. Among MYB TFs, R2R3-type MYB transcription factors are a vital group having a DNA binding site at the N-terminal and catalytic domain at the C-terminal. These TFs have a major function in signal transduction and secondary metabolite biosynthesis [45]. NtMYB12a and NtMYB12b both regulate the biosynthesis of flavonoids in tobacco, but NtMYB12a hinders the formation of fatty acids in tobacco seeds and leaves [46]. This R2R3-type MYB also negatively regulates the flavonoid biosynthesis pathway. In apples, MdMYB28 adversely controls anthocyanin biosynthesis. HY5 (elongated hypocotyl 5), a bZIP TF gene, is crucial for the biosynthesis of flavonoids and improves G. biloba's resistance to UV-B radiation. MdWRKY11 overexpression increases the accumulation of flavonoids and anthocyanin in apple Calli by promoting the expression of F3H, FLS, DFR, anthocyanidin synthase (ANS), and UFGT [47]. Other than genes and transcription factors, microRNA and lncRNA also play a vital role in flavonoid synthesis. Protein-coding genes are regulated by lncRNA, which takes part in the biosynthesis of G. biloba flavonoids [48]. A. thaliana plants with miPEP858a-edited and miPEP858a-overexpressing lines show changed plant development and flavonoid accumulation [49]. Flavonol concentrations in potatoes are raised by overexpression of MYB12A and MYB12C [50].

4. Physiological mechanism behind flavonoid-mediated drought stress tolerance

The primary mechanism by which flavonoids provide drought stress tolerance is their antioxidative function. Flavonoid biosynthesis is upregulated in specific cellular locales where ROS production is enriched. The possible sites of flavonoid accumulation are chloroplast, mitochondria, peroxisomes, and vacuole. Flavonoids have a variable play in mitigation, under drought where ROS scavenging, electron transport system maintenance, photosynthetic pigments protection, and stomatal regulation are involved [16]. These potentially detailed mechanisms of flavonoid actions are discussed below also depicted in **Figure 1**.

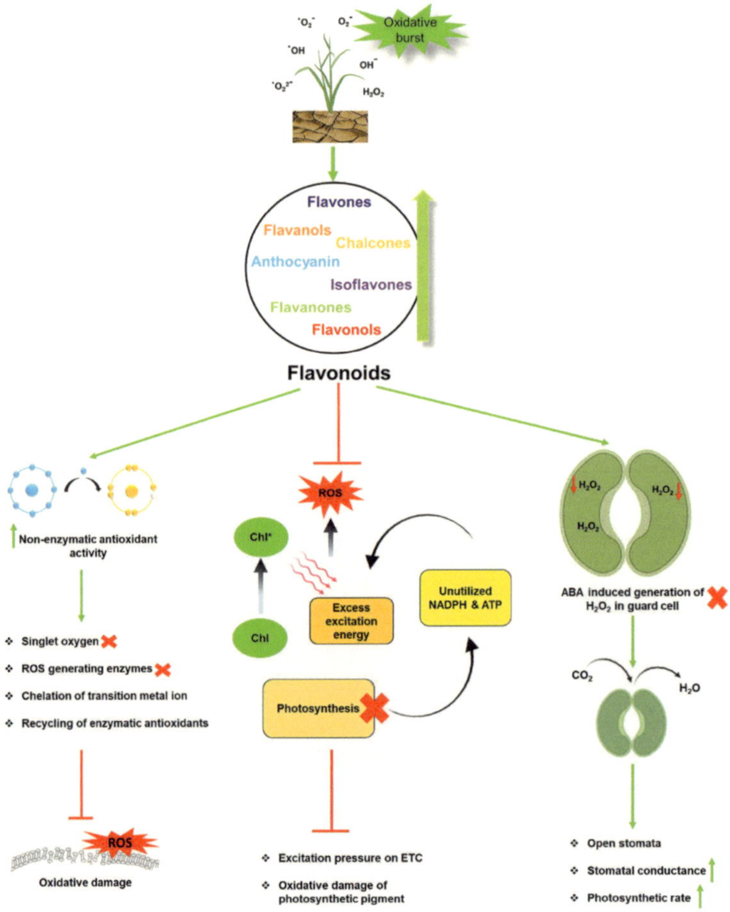

Figure 1.
Physiological mechanisms underlying the flavonoid-mediated drought stress tolerance.

Drought stress in plant is generally associated with the production of reactive oxygen species (ROS) which promotes oxidative stress in plant. Drought stress stimulate the production of flavonoids in plant. Flavonoids, due to their antioxidative nature prevent ROS generation by quenching singlet oxygen, suppress ROS generating enzyme activity, chelate transition metal ion, and recycle antioxidative enzymes. Thus protects the cell membrane and macromolecules from oxidation. Normally photosynthesis is hampered during drought stress. So, the NADPH and ATP produced in light reaction remain unutilized and it feedback regulates the operation of electron transport chain. In turn the excess excitation energy which is not photo quenched leads to production of ROS. Flavonoids are known to reduce the excitation pressure on ETC and also protect the photosynthetic pigments from oxidative damage. Flavonoids also inhibit the ABA-induce H_2O_2 production in the guard cell during drought stress. This results in open stomata and increased the stomatal conductance as well as photosynthetic rate even during drought stress condition.

4.1 Increased activity of enzymatic and non-enzymatic antioxidants

The role of flavonoids as antioxidants is well documented. A rise in the cellular pool of flavonoids can be seen in response to the oxidative stress caused by drought. The antioxidant role of flavonoids in scavenging superoxide radicals in *Ligustrum vulgare* leaf under drought stress has been reported by [51]. A markable accumulation of antioxidant flavonoids in leaf tissue was seen in response to drought stress. From the electronic paramagnetic resonance measurement, they observed that the quenching ability of flavonoids for O_2^- was in the order flavonol (quercetin 3-O-rutinoside) > dihydroxyflavone (luteolin 7-O-glucoside) > monohydroxyflavones (luteolin 4′-O-glucoside and apigenin 7-O-glucoside). It is reported that a positive correlation of total flavonoid content with traits related to photochemical quenching i.e., qL, qP, and Y(PSII) under drought stress conditions in lettuce seedlings [13]. Flavonoid accumulation pattern showed a quite similar trend with overall antioxidant activities estimated from FRAP and ABTS assays with a highly positive correlation, showing a significant contribution of flavonoids in antioxidant activities. The interconnected role of flavonoid content and antioxidant activities (measured by ABTS and DPPH method) under drought stress has been reported in *O. Africana* [11]. With the accumulation of flavonoids in drought stress the H_2O_2 content decreased giving drought stress tolerance in *Adonis* species [12]. According to [52], the tomato WD40 protein SlAN11 is essential for controlling flavonoid biosynthesis and seed dormancy through its coordination with bHLH and MYB proteins. The active role of catechin, epicatechin, and gallocatechin (flavan-3-ols) in neutralizing highly reactive free radicals and sustaining the physiological function of cells [53]. Fluvic acid strengthened ROS removal from tea plants under drought stress by stimulating the biosynthesis of kaempferol, quercetin, and myricetin [54]. It has been observed overaccumulation of kaempferol and quercetin in flavanone 3-hydroxylase overexpressing transgenic rice plants which due to strong non-enzymatic antioxidant activity caused a significant reduction in ROS production [10]. This reduced the impact of drought stress in transgenic plants than wild type which was evident from a lower level of salicylic acid and increased synthesis of dehydrin.

Flavonoids are polyphenolic compounds with multiple hydroxyl groups [16]. Their antioxidant ability arises due to their enhanced capacity to donate hydrogen or electrons. Flavonoids are oxidized by the ROS molecules in the chloroplast which in turn are recycled to reduced form by ascorbate. Under stress conditions, there is a dramatic increase in the content of ascorbate in the vacuole [15]. Flavonoids neutralize the harmful impact of ROS by (i) suppressing singlet oxygen, (ii) inhibiting the activity of ROS-generating enzymes such as xanthine oxidase, cyclooxygenase, monooxygenase, and lipoxygenase, (iii) chelation of transition metal ions such as iron and copper radicals, and (iv) recycling of other antioxidants [55]. Following these mechanisms, flavonoids quench the highly reactive ROS and prevent cellular peroxidation during drought stress.

4.2 Reduced excitation pressure on the ETC

Under drought stress, the plant closes its stomata to prevent server water loss. This hampers the entry and fixation of CO_2 due to which the NADPH and ATP produced through ETC remain unutilized [56]. In plants, there should be a balance in the light energy absorbed, excitation transfer, charge separation, electron transfer to the reaction center, generation of NADPH and ATP, and utilization of reducing power for CO_2 fixation [57]. The unused NADPH and ATP indicate that the captured light energy is more than the energy that can be used. This tends to regulate the electron transport rate [58, 59]. If the excess excitation energy is not quenched by photochemical reaction or heat, it leads to the generation of ROS, such as $^1O_2$, superoxide anion radical ($O_2^{\bullet -}$), and hydrogen peroxide (H_2O_2) [54]. This can damage the photosynthetic apparatus by photooxidative stress, where PSII is the potent target site [60].

Flavonoids due to their antioxidant activity are known to quench the ROS generated in chloroplast by excess excitation energy [61, 62]. The generation of excess excitation energy is highly linked with cell flavonoid deposition [63, 64]. Higher accumulation of flavonoids in the younger leaves of the Arabidopsis plant improved the PSII photochemistry under moderate drought stress. The excess excitation energy as well as excitation pressure was lower on these plants and they maintained a lower non-regulated energy dissipation by PSII which resulted in lower $^1O_2$ (ROS) generation. This reduced the membrane lipid peroxidation and maintained a healthy photosynthetic machinery. Ultimately increased flavonoid content resulted in acclimation of Arabidopsis plants to drought stress conditions [65].

4.3 Protection of photosynthetic pigments

The decline in the leaf's photosynthetic pigment content is a marker of oxidative stress in drought conditions due to the production of ROS [66, 67]. Photosynthetic pigments and enzymes are the main target sites for oxidation which makes photosynthesis highly sensitive to ROS [68]. There are several studies reported where drought stress affects the activity of both PSI and PSII. Along with that drought stress can also hamper photosynthesis by pigment degradation and by impairment of the pigment biosynthesis pathway [69]. The singlet oxygen produced in the chloroplast is efficiently quenched by the flavonoids which is essential to protect the thylakoid membrane as well as chloroplast outer envelope from lipid peroxidation [62]. Flavonoids help to maintain the H_2O_2 concentration at the sublethal level by acting as a substrate for peroxidases to reduce H_2O_2. Under extreme stress conditions, sometimes H_2O_2 escapes the chloroplast which is neutralized by the flavonoids located in the vacuole [15].

Flavonoids are involved in the protection of photosynthetic pigments owing to their antioxidant nature [20]. It has been reported that, in response to the decline in chlorophyll content and chlorophyll a/b ratio on the onset of drought, an increase in flavonoid and anthocyanin content was observed in *Pimpinella anisum* leaf [70]. This aligned with enhanced carotene and xanthophyll accumulation in the leaf suggesting a positive correlation with flavonoid content. This showed a protective role of flavonoids towards photosynthetic pigments. With the progression of drought stress, as the *Populus* plant accumulated H_2O_2 and O_2^- therefore was an increase in the deposition of flavonoids [68]. Owing to the antioxidative properties, the high flavonoid content helped in neutralizing the ROS. This prevented the oxidation of photosynthetic pigments and the plant was able to maintain higher chl a and chl b content even during the period of drought stress. In Refs., [67] highlighted that the transgenic rice plant overexpressing chalcone isomerase 2 exhibited increased resilience to drought stress as compared to the wild type. The transgenic lines showed greater RWC, reduced electrolyte leakage, and sustained more photosynthetic pigments and photosynthetic activity in drought conditions. Additionally, these lines reported elevated flavonoid content in response to drought, which suggested a potent role of flavonoids in photosynthetic pigment conservation. The transgenic tobacco lines overexpressing the *LcF3H* gene reported significant improvement in the content of flavan-3-ols including catechin, epicatcehin and epi-gallocatechin upon drought imposition. This improved the photosynthetic damage tolerance in the transgenic plant in drought stress. Moreover, the decline in the D1 protein content which is the main target for the photodamage in the PSII complex was less in transgenic plants [53]. Overexpression of CsF3′H in *Nicotiana benthamiana* transgenic plant stimulated the accumulation of eriodictoyl and dihydroquercetin. This improved the dehydration tolerance with noticeably higher chlorophyll content along with lesser MDA content in transgenic plant under drought stress [71].

4.4 Stomatal regulation

A direct impact of drought stress on plants is reduced stomatal aperture and mesophyll conductance which reduce the CO_2 availability in the guard cell. While the indirect effect of drought stress is related to alteration in photosynthetic activity [35]. ABA has a significant role in stomatal regulation during drought stress. ABA enhances the ROS accumulation in the guard cell. ABA together with ROS triggered the generation of H_2O_2 resulting in the activation of anion channel in the guard cell. This alters the osmolarity of the guard cell and drive stomatal closure [48]. Another mechanism by which flavonoids assist in drought tolerance is by regulating stomatal movement. Huge accumulation of flavonoids in the guard cell response to drought stress has been reported before.

Previously the role of flavonoids in drought stress tolerance in maize seedlings by regulating stomatal movements have been reported [72]. They observed a higher deposition of flavonols in the stomatal guard cells of *doi57* (*drought overly insensitivity*) mutant maize plant in drought stress as compared to the wild type. The mutants showed a faster water loss and lower leaf surface temperature owing to a comparatively open stomatal aperture along with higher stomatal conductance and transpiration rate even under drought stress conditions. This is accompanied by enhanced photosynthetic as well as water use efficiency in the mutants than the wild type. A similar role of flavonols in suppressing the buildup of ABA-induced H_2O_2 in the guard cell and ABA-triggered stomatal closure has been reported [73]. From a non-targeted metabolomics study on the guard cell of *Brassica napus* under drought stress induced by ABA treatment, [74] found upregulation in flavone and flavonol biosynthesis. They observed a reduction in ABA-induced stomatal closure response to 1 μM quercetin treatment in Arabidopsis. This indicates that quercetin may counteract ABA signaling probably by mitigating the ROS generated that contribute to stomata closure. The chalcone synthase mutant *tt4-2* of the Arabidopsis plant which is flawed in flavonol synthesis showed a hypersensitive response to ABA treatment and the extent of stomata closure was more pronounced than the wild type. According to [75], the deposition of flavonol in the guard cell of wild-type plants kept the stomatal aperture relatively open.

Another strategy to cope with drought stress is by closing of stomata to reduce transpirational water loss and sustain plant water balance [48]. ABA plays a major role in this process. Under severe water stress on the application of ABA, there is a significant decline in the width-to-length ratio of the stomatal aperture leading to stomatal closure in pigeon pea plants [14]. This aligned with the accumulation of flavonoids genistein, genistin and pterostilbene in the stomatal guard cell. In another perspective, this is one of the mechanisms exhibited by ABA to prevent severe water loss under drought stress by reducing stomatal aperture and by lowering the H^+ influx from the root tip in order to promote root growth. While studying two sea buckthorn species XY and FN, [76] observed a decline in leaf water potential, stomatal conductance, and photosynthetic rate under drought stress. Alongside the flavone, flavonol, isoflavone, and flavanone content depleted dramatically in the XY species making it more susceptible to drought stress. On the contrary, FN showed better tolerance to drought stress owing to a higher content of flavone and ABA. ABA-dependent signal transmission and flavonoid-dependent free radical scavenging collectively improved drought response in FN species. Due to their antioxidative property, flavonoids aid in maintaining the redox balance in guard cells by mitigating the ROS. Thus, flavonoids increase the responsiveness of stomatal guard cell and prevent stomata closure during drought stress [74]. Open stomata with greater mesophyll conductance is essential during drought stress to reduce leaf surface temperature and maintenance of photosynthetic efficiency.

5. Signal transduction mechanism of flavonoids under drought stress

5.1 Role of different flavonols and biosynthetic genes of flavanols under drought stress

Different enzymatic and non-enzymatic antioxidants play a major role in regulating the reactive oxygen species in drought-stress conditions. Among different non-enzymatic antioxidants flavonoids play a major role in regulating the concentration of different types of reactive oxygen species under drought stress. The flavonoid synthetic pathway is regulated by different hormones, including auxin, ABA, and ethylene [77]. Under drought stress, different flavonoid biosynthetic genes like dihydroflavonol-4-reductase, leuco anthocyanidin reductase, and leucoanthocyanidin dioxygenase are involved in synthesizing different types of flavonoids [78, 79] found that expression of phenylalanine-ammonia-lyase, cinnamic acid 4-hydroxylase, and 4-coumarate CoA ligase genes were highly upregulated in tea plants under drought stress. In the UGT76E11 transgenic line, flavonols act as signaling molecules to activate different other transcription factors under drought stress conditions [80]. In Arabidopsis, it was observed that over-expression of the VvbHLH1 gene could increase the concentration of flavonoids under drought-stress conditions [81]. Among different genes, chalcone isomerase 2 was highly expressed under drought stress. Other genes like OsCHI2 regulate the activity of different genes involved in the biosynthesis of different types of flavonoids. Rd29A::OsCHI12 transgenic rice plants showed higher photosynthetic activity under drought stress [67].

5.2 ABA-mediated signal transduction mechanism of flavonoids

ABA plays a major role in regulating different transcription factors [82]. Among different transcription factors, OsMYB48-1 transcription factor regulates the expression of different stress-responsive genes. Overexpression of OsMYB48-1 activates the OsPP2C68 transcription factor, which encodes protein phosphatase 2C activates under different stress responses [83]. The OsRK1 gene encodes protein kinase and also activates ABA signaling pathways [84]. Both OsPP2C68 and OsRK1 factors are regulated by ABA signaling pathways. Further, these transcription factors activate different LEA proteins like RAB21, OsLEA3, RAB16C, and RAB16D [85]. Expression of these genes could enhance the tolerance ability under drought-stress conditions.

6. Flavonoid-induced gene expression under drought stress

Plants have adopted diverse mechanisms to cope with harsh environmental conditions. Various stress-responsive genes also come into play for that [86]. Different transcription factors like WRKY, bZIP, ERF/DREBs, and MYB play a critical role in the biotic stress tolerance mechanism. Further genes related to different molecular mechanisms like signal transduction, osmolysis, ROS, and flavonoid biosynthesis pathway play an acritical role in abiotic stress tolerance mechanisms [87].

Flavonoids belong to the polyphenol group and actively take part in the plant defense mechanism by scavenging free radicals, ROS, thereby protecting complex molecules from damage that could impair cellular function, particularly under abiotic stress conditions [88]. These flavonoids are synthesized through the phenylpropanoid pathway. Different like *PAL*, *CHS*, *F3′H*, *DFR*, and *ANS are* very crucial for enhanced flavonoid concentration during drought conditions [89]. These genes are upregulated during drought stress.

In a study, it was found that anthocyanin structural genes DFR and ANS were found to be linked with salinity and ABA stress [90]. In transgenic potatoes, it was also observed that genes like CHS, CHI, and DFR enhance phenolics as well as anthocyanin levels due to abiotic stress [91]. Similarly in tobacco *EaCHS1 was* found to have a critical role in abiotic stress management through RNAi interference. [47], also proved on CHS gene in Arabidopsis under osmotic stress. Similarly, ectopic expression of the *F3H* gene from the halophyte *Lycium chinense* enhanced drought tolerance in transgenic tobacco [53].

In another study, the rice microarray data revealed that the CHI2 gene (LOC_Os06g10210) is upregulated under drought stress [67]. The gene was cloned from the drought-tolerant upland rice variety Nagina 22 (N22) and placed under the control of the stress-inducible AtRd29A promoter. To investigate its function, transgenic plants were developed in the drought-susceptible rice variety Pusa Sugandh 2 (PS2) and evaluated for drought tolerance. Furthermore, the gene's role in other abiotic stresses was assessed to gain a broader understanding of its function under different stress conditions.

Expression study of the genes in the wheat revealed that genes like *TaDFR*, *TaCHS*, *TaCHI*, *TaFNS*, *TaF3H*, *TaFLS*, and *TaANS* showed higher expression in Chinese spring compared to Aikang 58. While the *TaFNS* gene showed higher expression in drought stress. These findings suggest that the expression of flavonoid pathway genes and the accumulation of flavonoid compounds may play a pivotal role in wheat's drought tolerance and that the flavonoid response mechanisms may differ between cultivars [89].

In rice, genes related to flavonoid biosynthesis like *OsDfr* and *OsAns* also activated due to dehydration [90]. *RsF3H* gene also plays a crucial counterpart in stress tolerance by synthesizing flavonoid production in *Reaumuria soongorica*. Again, gene *CsCYT75B1* synthesizes more flavonoids with an enhanced antioxidant activity which ultimately enhances drought tolerance in *Arabido psis* [92]. Similarly, in *Fagopyrum leptopodum, FlbZIP12* genes modulated drought tolerance through the flavonoid biosynthesis pathway. In *Malus domestica,* a heat shock protein *MdHSFA8a* combined with chaperone HSP90 modulates drought stress through flavonoid accumulation [93]. Similarly, [94] in *Bupleurum chinense* and [95] in tea reported higher flavonoid accumulation under drought stress.

7. Conclusion

Flavonoids perform many functions to combat the negative impacts of drought, including acting as a primary protective agent for plants during water scarcity. The powerful antioxidative characteristics of these secondary metabolites assist in the degradation of oxidative stress by reactive oxygen species, thus boosting the plant's tolerability towards the stress condition. Furthermore, flavonoids are also important to the plant's survival during periods of drought due to their ability to modulate essential physiological and biochemical processes, such as signaling, the opening and closing of stomata, and cells' osmoregulation. Gaining knowledge on the biosynthesis and regulation of these flavonoids under drought stress provides valuable insights for developing stress-resilient crops. Therefore, these changes mark a positive direction for future progress in agricultural and biotechnological aspects.

Acknowledgements

Thanks are due to the Department of Crop Physiology and Biochemistry, ICAR-NRRI, for providing the necessary facilities.

Conflict of interest

The authors declare there is no conflict of interest.

Author details

Jajati Keshari Nayak, Dipsikha Mohanty, Debashis Mahapatra, Subhankar Mondal, Ashish Mohanty, Sushree Sangeeta, Gyanisha Nayak, Jeetendra Senapaty and Babyrani Panda*
ICAR-National Rice Research Institute, Cuttack, Odisha, India

*Address all correspondence to: babyranipanda94@gmail.com

References

[1] Zhang H, Zhu J, Gong Z, Zhu JK. Abiotic stress responses in plants. Nature Reviews Genetics. 2022;**23**:104-119

[2] Paparella F, D'Agostino DA, Burt J. Long-term, basin-scale salinity impacts from desalination in the Arabian/ Persian Gulf. Scientific Reports. 2022;**12**(1):20549

[3] Yadav S, Modi P, Dave A, Vijapura A, Patel D, Patel M. Effect of abiotic stress on crops. Sustainable Crop Production. 2020;**3**:5-16

[4] Daryanto S, Wang L, Jacinthe PA. Global synthesis of drought effects on maize and wheat production. PLoS One. 2016;**11**:e0156362

[5] Farooq M, Gogoi N, Barthakur S, Baroowa B, Bharadwaj N, Alghamdi SS, et al. Drought stress in grain legumes during reproduction and grain filling. Journal of Agronomy and Crop Science. 2017;**203**:81-102

[6] Bhandari U, Gajurel A, Khadka B, Thapa I, Chand I, Bhatta D, et al. Morpho-physiological and biochemical response of rice (*Oryza sativa* L.) to drought stress: A review. Heliyon. 2023;**9**(3):e13744

[7] Fang Y, Xiong L. General mechanisms of drought response and their application in drought resistance improvement in plants. Cellular and Molecular Life Sciences. 2015;**72**:673-689

[8] Lum MS, Hanafi MM, Rafii YM, Akmar AS. Effect of drought stress on growth, proline and antioxidant enzyme activities of upland rice. JAPS: Journal of Animal & Plant Sciences. 2014;**24**(5):1487-1493

[9] Boogar AR, Salehi H, Jowkar A. Exogenous nitric oxide alleviates oxidative damage in turfgrasses under drought stress. South African Journal of Botany. 2014;**92**:78-82

[10] Jan R, Khan MA, Asaf S, Lubna WM, Park JR, Asif S, et al. Drought and UV radiation stress tolerance in rice is improved by overaccumulation of non-enzymatic antioxidant flavonoids. Antioxidants. 2022;**11**:917

[11] Talbi S, Rojas JA, Sahrawy M, Rodríguez-Serrano M, Cárdenas KE, Debouba M, et al. Effect of drought on growth, photosynthesis and total antioxidant capacity of the saharan plant Oudeneya africana. Environmental and Experimental Botany. 2020;**176**:104099

[12] Gao S, Wang Y, Yu S, Huang Y, Liu H, Chen W, et al. Effects of drought stress on growth, physiology and secondary metabolites of two Adonis species in Northeast China. Scientia Horticulturae. 2020;**259**:108795

[13] Shin YK, Bhandari SR, Jo JS, Song JW, Lee JG. Effect of drought stress on chlorophyll fluorescence parameters, phytochemical contents, and antioxidant activities in lettuce seedlings. Horticulturae. 2021;7:238

[14] Yang W, Li N, Fan Y, Dong B, Song Z, Cao H, et al. Transcriptome analysis reveals abscisic acid enhancing drought resistance by regulating genes related to flavonoid metabolism in pigeon pea. Environmental and Experimental Botany. 2021;**191**:104627

[15] Baskar V, Venkatesh R, Ramalingam S. Flavonoids (antioxidants systems) in higher plants and their response to stresses. In: Gupta D,

Palma J, Corpas F. editors. Antioxidants and Antioxidant Enzymes in Higher Plants. Cham: Springer; 2018. pp. 253-268

[16] Shomali A, Das S, Arif N, Sarraf M, Zahra N, Yadav V, et al. Diverse physiological roles of flavonoids in plant environmental stress responses and tolerance. Plants. 2022;**11**:3158

[17] Mathesius U. Flavonoid functions in plants and their interactions with other organisms. Plants. 2018;**7**:30

[18] Nakabayashi R, Yonekura-Sakakibara K, Urano K, Suzuki M, Yamada Y, Nishizawa T, et al. Enhancement of oxidative and drought tolerance in Arabidopsis by overaccumulation of antioxidant flavonoids. The Plant Journal. 2014;**77**:367-379

[19] Di Ferdinando M, Brunetti C, Fini A, Tattini M. Flavonoids as antioxidants in plants under abiotic stresses. In: Ahmad P, Prasad M, editors. Abiotic Stress Responses in Plants. New York: Springer; 2012. pp. 159-179

[20] Agati G, Azzarello E, Pollastri S, Tattini M. Flavonoids as antioxidants in plants: Location and functional significance. Plant Science. 2012;**196**:67-76

[21] Guven H, Arici A, Simsek O. Flavonoids in our foods: A short review. Journal of Basic and Clinical Health Sciences. 2019;**3**:96-106

[22] Sánchez M, Romero M, Gómez-Guzmán M, Tamargo J, Pérez-Vizcaino F, Duarte J. Cardiovascular effects of flavonoids. Current Medicinal Chemistry. 2019;**26**:6991-7034

[23] Pan F, Liu Y, Liu J, Wang E. Stability of blueberry anthocyanin, anthocyanidin and pyranoanthocyanidin pigments and their inhibitory effects and mechanisms in human cervical cancer HeLa cells. RSC Advances. 2019;**9**:10842-10853

[24] Tena N, Martín J, Asuero AG. State of the art of anthocyanins: Antioxidant activity, sources, bioavailability, and therapeutic effect in human health. Antioxidants. 2020;**23**(9):451

[25] Houghton A, Appelhagen I, Martin C. Natural blues: Structure meets function in anthocyanins. Plants. 2021;**10**:726

[26] Iorizzo M, Curaba J, Pottorff M, Ferruzzi MG, Simon P, Cavagnaro PF. Carrot anthocyanins genetics and genomics: Status and perspectives to improve its application for the food colorant industry. Genes. 2020;**11**:906

[27] Soares JC, Rosalen PL, Lazarini JG, Massarioli AP, da Silva CF, Nani BD, et al. Comprehensive characterization of bioactive phenols from new Brazilian superfruits by LC-ESI-QTOF-MS, and their ROS and RNS scavenging effects and anti-inflammatory activity. Food Chemistry. 2019;**281**:178-188

[28] Khoo HE, Azlan A, Tang ST, Lim SM. Anthocyanidins and anthocyanins: Colored pigments as food, pharmaceutical ingredients, and the potential health benefits. Food & Nutrition Research. 2017;**61**(1):1361779

[29] Chen Q, Wang D, Tan C, Hu Y, Sundararajan B, Zhou Z. Profiling of flavonoid and antioxidant activity of fruit tissues from 27 Chinese local citrus cultivars. Plants. 2020;**5**(9):196

[30] Liu L, Zuo Z, Lu S, Liu A, Liu X. Naringin attenuates diabetic retinopathy by inhibiting inflammation, oxidative stress and NF-κB activation in vivo and

in vitro. Iranian Journal of Basic Medical Sciences. 2017;**20**:813

[31] Hollósy F. Effects of ultraviolet radiation on plant cells. Micron. 2002;**33**(2):179-197

[32] Danciu C, Avram S, Pavel IZ, Ghiulai R, Dehelean CA, Ersilia A, et al. Main isoflavones found in dietary sources as natural anti-inflammatory agents. Current Drug Targets. 2018;**19**:841-853

[33] Alshehri MM, Sharifi-Rad J, Herrera-Bravo J, Jara EL, Salazar LA, Kregiel D, et al. Therapeutic potential of isoflavones with an emphasis on daidzein. Oxidative Medicine and Cellular Longevity. 2021;**2021**:6331630

[34] Canivenc-Lavier MC, Bennetau-Pelissero C. Phytoestrogens and health effects. Nutrients. 2023;**15**:317

[35] Wang M, Firrman J, Liu L, Yam K. A review on flavonoid apigenin: Dietary intake, ADME, antimicrobial effects, and interactions with human gut microbiota. BioMed Research International. 2019;**2019**:7010467

[36] Zhou K, Yang S, Li SM. Naturally occurring prenylated chalcones from plants: Structural diversity, distribution, activities and biosynthesis. Natural Product Reports. 2021;**38**:2236-2260

[37] Jasim HA, Nahar L, Jasim MA, Moore SA, Ritchie KJ, Sarker SD. Chalcones: Synthetic chemistry follows where nature leads. Biomolecules. 2021;**11**(8):1203

[38] Lin J, Lai G, Guo A, He L, Yang F, Huang Y, et al. Overexpression of LAR1 suppresses anthocyanin biosynthesis by enhancing catechin competition leading to promotion of proanthocyanidin pathway in spine grape (*Vitis davidii*) cells. International Journal of Molecular Sciences. 2024;**25**:12087

[39] Khan RA, Abbas N. Role of epigenetic and post-translational modifications in anthocyanin biosynthesis: A review. Gene. 15 Dec 2023;**887**:147694

[40] Liu W, Feng Y, Yu S, Fan Z, Li X, Li J, et al. The flavonoid biosynthesis network in plants. International Journal of Molecular Sciences. 2021;**22**:12824

[41] Zheng J, Sun R, Wu D, Chen P, Zheng P. Engineered Zea mays phenylalanine ammonia-lyase for improve the catalytic efficiency of biosynthesis trans-cinnamic acid and p-coumaric acid. Enzyme and Microbial Technology. 2024;**176**:110423

[42] Martín JF, Liras P. Comparative molecular mechanisms of biosynthesis of naringenin and related chalcones in actinobacteria and plants: Relevance for the obtention of potent bioactive metabolites. Antibiotics. 2022;**11**:82

[43] Ku YS, Ng MS, Cheng SS, Lo AW, Xiao Z, Shin TS, et al. Understanding the composition, biosynthesis, accumulation and transport of flavonoids in crops for the promotion of crops as healthy sources of flavonoids for human consumption. Nutrients. 2020;**12**:1717

[44] Zhai R, Zhao Y, Wu M, Yang J, Li X, Liu H, et al. The MYB transcription factor PbMYB12b positively regulates flavonol biosynthesis in pear fruit. BMC Plant Biology. 2019;**19**:1-1

[45] Qin W, Xie L, Li Y, Liu H, Fu X, Chen T, et al. An R2R3-MYB transcription factor positively regulates the glandular secretory trichome initiation in Artemisia annua L. Frontiers in Plant Science. 2021;**12**:657156

[46] Wang Z, Li ZF, Wang SS, Xiao YS, Xie XD, Wu MZ, et al. NtMYB12a acts downstream of sucrose to inhibit fatty acid accumulation by targeting lipoxygenase and SFAR genes in tobacco. Plant, Cell & Environment. 2021;**44**:775-791

[47] Wang N, Liu W, Zhang T, Jiang S, Xu H, Wang Y, et al. Transcriptomic analysis of red-fleshed apples reveals the novel role of MdWRKY11 in flavonoid and anthocyanin biosynthesis. Journal of Agricultural and Food Chemistry. 2018;**66**:7076-7086

[48] Liu S, Wang L, Cao M, Pang S, Li W, Kato-Noguchi H, et al. Identification and characterization of long non-coding RNAs regulating flavonoid biosynthesis in Ginkgo biloba leaves. Industrial Crops and Products. 2020;**158**:112980

[49] Sharma A, Badola PK, Bhatia C, Sharma D, Trivedi PK. Primary transcript of miR858 encodes regulatory peptide and controls flavonoid biosynthesis and development in Arabidopsis. Nature Plants. 2020;**6**(10):1262-1274

[50] Lin S, Singh RK, Moehninsi M, Navarre DA. R2R3-MYB transcription factors, StmiR858 and sucrose mediate potato flavonol biosynthesis. Horticulture Research. 2021;**1**:8

[51] Tattini M, Galardi C, Pinelli P, Massai R, Remorini D, Agati G. Differential accumulation of flavonoids and hydroxycinnamates in leaves of *Ligustrum vulgare* under excess light and drought stress. New Phytologist. 2004;**163**:547-561

[52] Gao Y, Liu J, Chen Y, Tang H, Wang Y, He Y, et al. Tomato SlAN11 regulates flavonoid biosynthesis and seed dormancy by interaction with bHLH proteins but not with MYB proteins. Horticulture Research. 2018;**5**:1-18

[53] Song X, Diao J, Ji J, Wang G, Guan C, Jin C, et al. Molecular cloning and identification of a flavanone 3-hydroxylase gene from Lycium chinense, and its overexpression enhances drought stress in tobacco. Plant Physiology and Biochemistry. 2016;**98**:89-100

[54] Sun H, Shi Q, Zhang SB, Huang W. Coordination of cyclic electron flow and water–water cycle facilitates photoprotection under fluctuating light and temperature stress in the epiphytic orchid Dendrobium officinale. Plants. 2021;**10**:606

[55] Mierziak J, Kostyn K, Kulma A. Flavonoids as important molecules of plant interactions with the environment. Molecules. 2014;**19**:16240-16265

[56] Lawlor DW, Tezara W. Causes of decreased photosynthetic rate and metabolic capacity in water-deficient leaf cells: A critical evaluation of mechanisms and integration of processes. Annals of Botany. 2009;**103**:561-579

[57] McKew BA, Davey P, Finch SJ, Hopkins J, Lefebvre SC, Metodiev MV, et al. The trade-off between the light-harvesting and photoprotective functions of fucoxanthin-chlorophyll proteins dominates light acclimation in Emiliania huxleyi (clone CCMP 1516). New Phytologist. 2013;**200**:74-85

[58] Moustakas M, Bayçu G, Sperdouli I, Eroğlu H, Eleftheriou EP. Arbuscular mycorrhizal symbiosis enhances photosynthesis in the medicinal herb Salvia fruticosa by improving photosystem II photochemistry. Plants. 2020;**9**:962

[59] Zavafer A, Mancilla C. Concepts of photochemical damage of photosystem

II and the role of excessive excitation. Journal of Photochemistry and Photobiology C: Photochemistry Reviews. 2021;**47**:100421

[60] Moustakas M, Sperdouli I, Kouna T, Antonopoulou CI, Therios I. Exogenous proline induces soluble sugar accumulation and alleviates drought stress effects on photosystem II functioning of *Arabidopsis thaliana* leaves. Plant Growth Regulation. 2011;**65**:315-325

[61] Majer P, Neugart S, Krumbein A, Schreiner M, Hideg É. Singlet oxygen scavenging by leaf flavonoids contributes to sunlight acclimation in *Tilia platyphyllos*. Environmental and Experimental Botany. 2014;**100**:1-9

[62] Dall'Osto L, Cazzaniga S, Bressan M, Paleček D, Židek K, Niyogi KK, et al. Two mechanisms for dissipation of excess light in monomeric and trimeric light-harvesting complexes. Nature Plants. 2017;**3**:1-9

[63] Sumbele S, Fotelli MN, Nikolopoulos D, Tooulakou G, Liakoura V, Liakopoulos G, et al. Photosynthetic capacity is negatively correlated with the concentration of leaf phenolic compounds across a range of different species. AoB Plants. 1 Jan 2012;**2012**:pls025

[64] Gog L, Zavala J, DeLucia EH. Linking primary and secondary metabolism a mechanistic hypothesis for how elevated CO_2 modulates defenses. In: Gog L, Zavala J, DeLucia EH, editors. The Biology of Plant-Insect Interactions. Florida: CRC Press; 2018. pp. 93-112

[65] Sperdouli I, Moustaka J, Ouzounidou G, Moustakas M. Leaf age-dependent photosystem II photochemistry and oxidative stress responses to drought stress in Arabidopsis thaliana are modulated by flavonoid accumulation. Molecules. 2021;**26**:4157

[66] Rai AC, Singh M, Shah K. Engineering drought tolerant tomato plants over-expressing BcZAT12 gene encoding a C_2H_2 zinc finger transcription factor. Phytochemistry. 2013;**85**:44-50

[67] Jayaraman K, Sevanthi AM, Sivakumar SR, Viswanathan C, Mohapatra T, Mandal PK. Stress-inducible expression of chalcone isomerase2 gene improves accumulation of flavonoids and imparts enhanced abiotic stress tolerance to rice. Environmental and Experimental Botany. 2021;**190**:104582

[68] Ahmed U, Rao MJ, Qi C, Xie Q, Noushahi HA, Yaseen M, et al. Expression profiling of flavonoid biosynthesis genes and secondary metabolites accumulation in populus under drought stress. Molecules. 2021;**26**(18):5546

[69] Ashraf MH, Harris PJ. Photosynthesis under stressful environments: An overview. Photosynthetica. 2013;**51**:163-190

[70] Asadi Kavan Z, Ghorbanli M, Sateei A. The effect of drought stress and exogenous ascorbate on photosynthetic pigments, flavonoids, phenol compounds and lipid peroxidation in Pimpinella anisum L. Iranian Journal of Medicinal and Aromatic Plants Research. 2010;**25**(4):456-469

[71] Baba SA, Ashraf N. Functional characterization of flavonoid 3′-hydroxylase, CsF3′ H, from *Crocus sativus* L: Insights into substrate specificity and role in abiotic stress. Archives of Biochemistry and Biophysics. 2019;**667**:70-78

[72] Li B, Fan R, Sun G, Sun T, Fan Y, Bai S, et al. Flavonoids improve drought tolerance of maize seedlings by regulating the homeostasis of reactive oxygen species. Plant and Soil. 2021;**461**:389-405

[73] Watkins JM, Chapman JM, Muday GK. Abscisic acid-induced reactive oxygen species are modulated by flavonols to control stomata aperture. Plant Physiology. 2017;**175**:1807-1825

[74] Zhu M, Assmann SM. Metabolic signatures in response to abscisic acid (ABA) treatment in Brassica napus guard cells revealed by metabolomics. Scientific Reports. 2017;**7**:12875

[75] Watkins JM, Hechler PJ, Muday GK. Ethylene-induced flavonol accumulation in guard cells suppresses reactive oxygen species and moderates stomatal aperture. Plant Physiology. 2014;**164**(4):1707-1717

[76] Gao G, Lv Z, Zhang G, Li J, Zhang J, He C. An ABA–flavonoid relationship contributes to the differences in drought resistance between different sea buckthorn subspecies. Tree Physiology. 2021;**41**:744-755

[77] Wu P, Liu A, Li L. Metabolomics and transcriptome analysis of the biosynthesis mechanism of flavonoids in the seeds of Euryale ferox Salisb at different developmental stages. Molecular Genetics and Genomics. 2021;**296**:953-970

[78] Upadhyaya H, Panda SK. Responses of Camellia sinensis to drought and rehydration. Biologia Plantarum. 2004;**48**:597-600

[79] Gai Z, Wang YU, Ding Y, Qian W, Qiu C, Xie H, et al. Exogenous abscisic acid induces the lipid and flavonoid metabolism of tea plants under drought stress. Scientific Reports. 2020;**10**(1):12275

[80] Li Q, Yu HM, Meng XF, Lin JS, Li YJ, Hou BK. Ectopic expression of glycosyltransferase UGT 76E11 increases flavonoid accumulation and enhances abiotic stress tolerance in Arabidopsis. Plant Biology. 2018;**20**(1):10-19

[81] Wang F, Zhu H, Chen D, Li Z, Peng R, Yao Q. A grape bHLH transcription factor gene, VvbHLH1, increases the accumulation of flavonoids and enhances salt and drought tolerance in transgenic Arabidopsis thaliana. Plant Cell, Tissue and Organ Culture (PCTOC). 2016;**125**:387-398

[82] Xiong H, Li J, Liu P, Duan J, Zhao Y, Guo X, et al. Overexpression of OsMYB48-1, a novel MYB-related transcription factor, enhances drought and salinity tolerance in rice. PLoS One. 2014;**9**(3):e92913

[83] Xue T, Wang D, Zhang S, Ehlting J, Ni F, Jakab S, et al. Genome-wide and expression analysis of protein phosphatase 2C in rice and Arabidopsis. BMC Genomics. 2008;**9**:1-21

[84] Chae MJ, Lee JS, Nam MH, Cho K, Hong JY, Yi SA, et al. A rice dehydration-inducible SNF1-related protein kinase 2 phosphorylates an abscisic acid responsive element-binding factor and associates with ABA signaling. Plant Molecular Biology. 2007;**63**:151-169

[85] RoyChoudhury A, Roy C, Sengupta DN. Transgenic tobacco plants overexpressing the heterologous lea gene Rab16A from rice during high salt and water deficit display enhanced tolerance to salinity stress. Plant Cell Reports. 2007;**26**:1839-1859

[86] Devi E, Kumar S, Basanta Singh T, Sharma SK, Beemrote A,

Devi CP, et al. Adaptation strategies and defence mechanisms of plants during environmental stress. In: Ghorbanpour M, Varma A, editors. Medicinal plants and environmental challenges. Cham: Springer; 2017. pp. 359-413

[87] Praveen A, Dubey S, Singh S, Sharma VK. Abiotic stress tolerance in plants: A fascinating action of defense mechanisms. 3 Biotech. 2023;**13**:102

[88] Kranner I, Birtić S. A modulating role for antioxidants in desiccation tolerance. Integrative and Comparative Biology. 2005;**45**(5):734-740

[89] Ma D, Sun D, Wang C, Li Y, Guo T. Expression of flavonoid biosynthesis genes and accumulation of flavonoid in wheat leaves in response to drought stress. Plant Physiology and Biochemistry. 2014;**80**:60-66

[90] Ithal N, Reddy AR. Rice flavonoid pathway genes, OsDfr and OsAns, are induced by dehydration, high salt and ABA, and contain stress responsive promoter elements that interact with the transcription activator, OsC1-MYB. Plant Science. 2004;**166**(6):1505-1513

[91] Łukaszewicz M, Szopa J. Pleiotropic effect of flavonoid biosynthesis manipulation in transgenic potato plants. Acta Physiologiae Plantarum. 2005;**27**:221-228

[92] Rao MJ, Xu Y, Tang X, Huang Y, Liu J, Deng X, et al. CsCYT75B1, a citrus CYTOCHROME P450 gene, is involved in accumulation of antioxidant flavonoids and induces drought tolerance in transgenic Arabidopsis. Antioxidants. 2020;**9**(2):161

[93] Wang N, Liu W, Yu L, Guo Z, Chen Z, Jiang S, et al. HEAT SHOCK FACTOR A8a modulates flavonoid synthesis and drought tolerance. Plant Physiology. 2020;**184**:1273-1290

[94] Yang LL, Yang L, Yang X, Zhang T, Lan YM, Zhao Y, et al. Drought stress induces biosynthesis of flavonoids in leaves and saikosaponins in roots of Bupleurum chinense DC. Phytochemistry. 2020;**177**:112434

[95] Gu H, Wang Y, Xie H, Qiu C, Zhang S, Xiao J, et al. Drought stress triggers proteomic changes involving lignin, flavonoids and fatty acids in tea plants. Scientific Reports. 2020;**10**(1):15504

Chapter 6

Perspective Chapter: From Defense to Development – Exploring Plant Secondary Metabolites

Archana Watts, Yuvonne Angel Lyngdoh, Anshul Watts and Shruti Sinha

Abstract

Plant secondary metabolites, often regarded as the unsung heroes of the botanical world, play pivotal roles in both defense mechanisms and developmental processes. Initially, these compounds evolved primarily as a means of protecting plants against herbivores, pathogens, and environmental stresses. Alkaloids, flavonoids, and terpenoids, for instance, serve as toxic deterrents, antimicrobial agents, or UV protectants. However, beyond their defensive roles, these metabolites are crucial in plant development, influencing processes such as growth regulation, pigment production, and symbiotic relationships. Recent research highlights how these versatile compounds contribute to the plant's adaptability and evolutionary success, offering a treasure trove of potential applications in agriculture, medicine, and biotechnology.

Keywords: defense, plant development, growth regulation, plant secondary metabolites, biotechnological applications

1. Introduction

Plant secondary metabolites (SM) are phytochemicals that are synthesized in small quantities in plants. They are not directly required for the normal growth and development of plants but are necessary for survival, ecological interactions, and adaptation of the plant under stress conditions. Unlike primary metabolites, which are required for normal growth and development of plants and are synthesized *via* primary metabolic pathways, secondary metabolites are synthesized using the byproducts of primary metabolic pathways, and secondary metabolic pathways are involved in the synthesis of SM.

Secondary metabolites are broadly classified into three major groups: phenolics, alkaloids, and terpenoids. Some compounds, such as alkaloids, flavonoids, antibiotics, and pigments, are unique to certain plant species. While often considered byproducts of primary metabolism, these molecules play vital roles in shaping plant ecology, species diversity, and survival strategies. Due to their specific chemical properties, secondary metabolites help plants combat various biotic and abiotic stresses, enabling them to grow and thrive in challenging environments. Some of these compounds are highly toxic, and capable of deterring or even killing herbivores and insects that feed on them. Interestingly, in some cases, herbivores ingest these toxic metabolites and convert them into nontoxic forms as a defense mechanism against their own predators. Additionally, the unique chemical compounds secreted by plant roots influence and regulate the surrounding soil microbiota and the growth of neighboring plants. In ecosystems, secondary metabolites are an essential component as they affect environmental equilibrium, plant–plant, plant-microbe, and plant-fungus interactions, and survival. The pigments, colors, nectar components, and alkaloids present in the plants help in attracting pollinators and thus help in the dispersal of seeds to distant places and distribute the seeds in diverse ecosystems. They also have a significant impact on microbial populations and nitrogen cycling, which are essential for soil health. These diverse functions enhance a plant's ability to survive and adapt. The remarkable variety of secondary metabolites across species contributes to specialized roles such as antimicrobial resistance, plant defense, and pigmentation [1]. Owing to their unique bioactive properties, secondary metabolites are widely utilized across various sectors, including agriculture (as biopesticides), medicine (as antibiotics and anticancer agents), cosmetics (in the form of essential oils), and food production (as flavoring and coloring agents). The dynamic and complex nature of these compounds reflects the intricate survival strategies plants have evolved over time, strategies that are far more sophisticated than previously understood. This complexity highlights the need for deeper exploration and study to fully unravel their ecological roles and potential applications.

This chapter will highlight the dual roles of secondary metabolites in both plant development and defense responses, emphasizing their significance in plant physiology and their contribution to the adaptation of plants within diverse ecosystems. The chapter will also focus on the biosynthetic pathways and developmental regulation of secondary metabolite production, along with their roles in plant defense, including stress tolerance, allelopathic interactions, and chemical defenses against herbivores and pathogens.

2. Biosynthesis and development

The use of secondary metabolites dates back to ancient history, when the plant extracts were used to kill enemies in wars or as poisons. Historically, plant species known for their medicinal or toxic properties have attracted considerable scientific interest. The exploration of such plants led to the isolation and characterization of some of the earliest plant-derived therapeutic compounds, including salicylic acid (aspirin), morphine, digitoxin, and quinine [2–5]. These efforts marked a turning point in natural product research, establishing the foundation for modern pharmacognosy. Subsequent studies focusing on the chemical constituents of medicinal plants facilitated the discovery of a wide range of secondary metabolites such as alkaloids, glycosides, anthocyanins, carotenoids, and terpenes, many of which are now recognized for their diverse biological activities and therapeutic potential [6].

2.1 Terpenoids

Terpenoids, also known as isoprenoids, represent one of the largest classes of natural products, with over 55,000 known compounds [7]. These secondary metabolites play crucial roles in plant physiology, including defense mechanisms, growth regulation, and interaction with the environment [8]. Terpenoids are biosynthesized from simple five-carbon precursors, isopentenyl pyrophosphate (IPP) and dimethylallyl pyrophosphate (DMAPP), through two major biosynthetic pathways: the mevalonate (MVA) pathway (**Figure 1**) and the methylerythritol phosphate (MEP) pathway (**Figure 2**) [9].

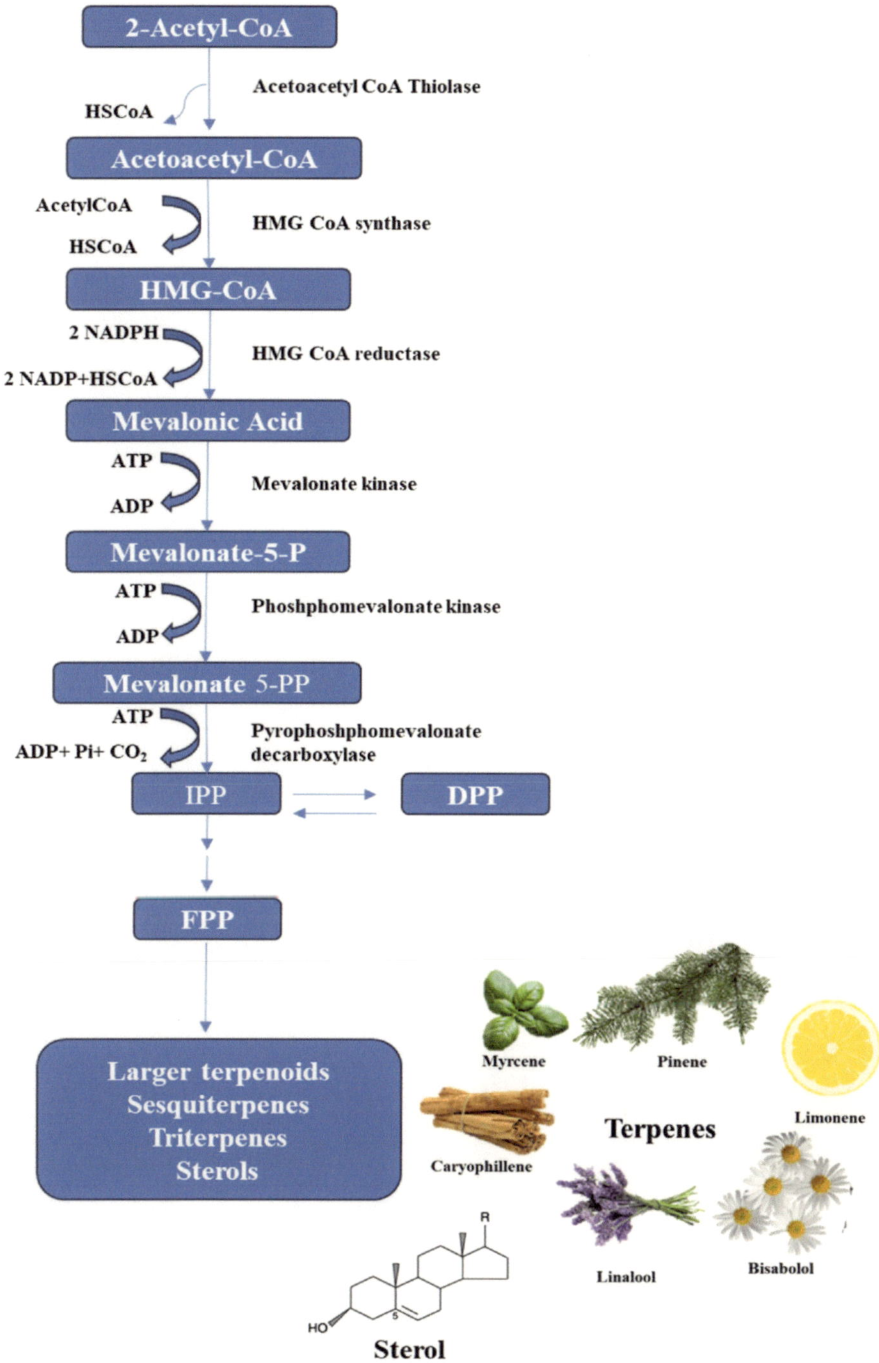

Figure 1.
Mevalonate (MVA) pathway: Synthesis of sterols, sesquiterpenes, and triterpenes. Acetyl-CoA (acetyl coenzyme A); Acetoacetyl-CoA (acetoacetyl coenzyme A); HMG-CoA (3-hydroxy-3-methylglutaryl coenzyme A); Mevalonate-5-P (mevalonate phosphate); Mevalonate-5-PP (mevalonate diphosphate); Isopentenyl-PP (isopentenyl diphosphate); DPP (dimethylallyl diphosphate); Farnesyl-PP (farnesyl diphosphate).

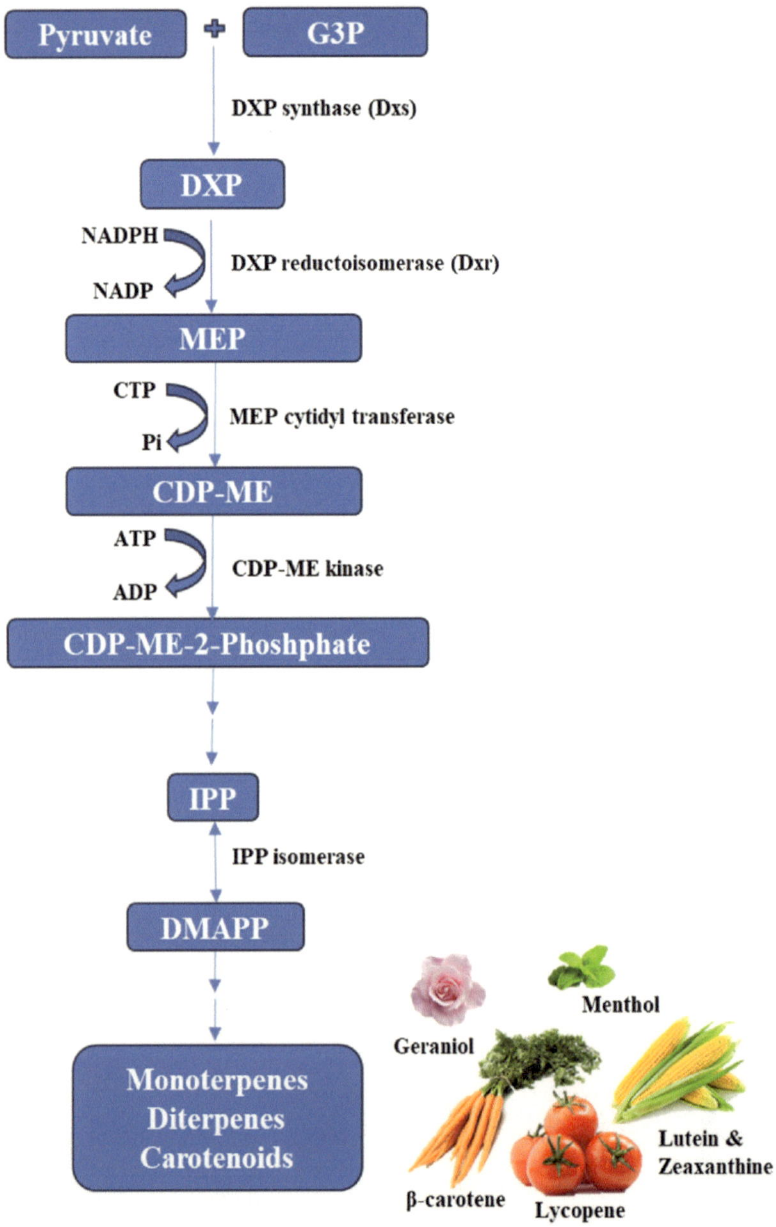

Figure 2.
MEP pathway: Synthesis of monoterpenes, diterpenes, and carotenoids. G3P (glyceraldehyde-3-phosphate); DXP (1-deoxy-D-xylulose-5-phosphate); MEP (2-C-methyl-D-erythritol 4-phosphate); CDP-ME (4-diphosphocytidyl-2-C-methyl-d-erythritol); IPP (isopentenyl diphosphate); DMAPP (dimethylallyl diphosphate).

3. Biosynthetic pathways of terpenoids

3.1 Mevalonate (MVA) pathway (cytosolic pathway)

The MVA pathway operates in the cytosol and is primarily responsible for the biosynthesis of sterols, sesquiterpenes (C15), and triterpenes (C30). The mevalonate (MVA) pathway initiates with acetyl-CoA as the primary substrate. In this pathway, two molecules of acetyl-CoA undergo a series of enzymatic reactions to ultimately produce isopentenyl diphosphate (IPP) and dimethylallyl diphosphate (DPP), which serve as essential isoprene units. A key regulatory step in this pathway is catalyzed by 3-hydroxy-3-methylglutaryl-CoA reductase (HMG-CoA reductase), the rate-limiting enzyme and a well-established pharmacological target for cholesterol-lowering therapies. Subsequently, IPP is converted to geranyl diphosphate (GPP), which further condenses to form farnesyl diphosphate (FPP). FPP acts as a central precursor for the biosynthesis of a wide array of isoprenoid compounds, including larger terpenoids, sesquiterpenes, triterpenes, and sterols.

3.2 Methylerythritol Phosphate (MEP) pathway (plastidial pathway)

The MEP pathway, occurring in the plastids, is responsible for the synthesis of monoterpenes (C10), diterpenes (C20), and carotenoids (C40). The methylerythritol phosphate (MEP) pathway serves as an alternative to the mevalonate (MVA) pathway and operates predominantly in most bacteria and plastids for the biosynthesis of isoprenoid units, including terpenes and carotenoids. This pathway begins with the condensation of pyruvate and glyceraldehyde-3-phosphate (G3P) to form 1-deoxy-D-xylulose-5-phosphate (DXP). DXP is subsequently converted to 2-C-methyl-D-erythritol 4-phosphate (MEP) through the action of the enzyme DXP reductoisomerase. Through a series of downstream enzymatic steps, MEP is ultimately transformed into isopentenyl diphosphate (IPP) and dimethylallyl diphosphate (DMAPP). These isoprene units act as precursors for the synthesis of various classes of terpenoids, including monoterpenes, diterpenes, and carotenoids.

4. Terpenoid classification

Isoprene is the building block of all classes of terpenes, wherein they are fused together in a head-to-tail fashion. Different classes of terpenes are thus defined based on the number of fused isoprene units as described in **Table 1**.

Isopentenyl pyrophosphate (IPP) and dimethylallyl pyrophosphate (DMAPP) condense to form geranyl pyrophosphate (GPP), which serves as a precursor for the synthesis of monoterpenes. When GPP combines with another IPP unit, it produces farnesyl pyrophosphate (FPP), the precursor for sesquiterpenes and triterpenes. Further addition of an IPP unit to FPP forms geranylgeranyl pyrophosphate (GGPP), which acts as a precursor for the biosynthesis of diterpenes and tetraterpenes (**Figure 3**).

Isoprene units	Examples
Monoterpenes (C10)	Limonene, Geraniol
Sesquiterpenes (C15)	Artemisinin, Farnesene
Diterpenes (C20)	Gibberellins, Taxol
Triterpenes (C30)	Squalene, Betulinic acid
Tetraterpenes (C40)	Carotenoids (Lycopene, β-Carotene)
Polyterpenes (>C40)	Natural rubber

Table 1.
Classification of terpenoids based on the number of isoprene (C5) units in their structure.

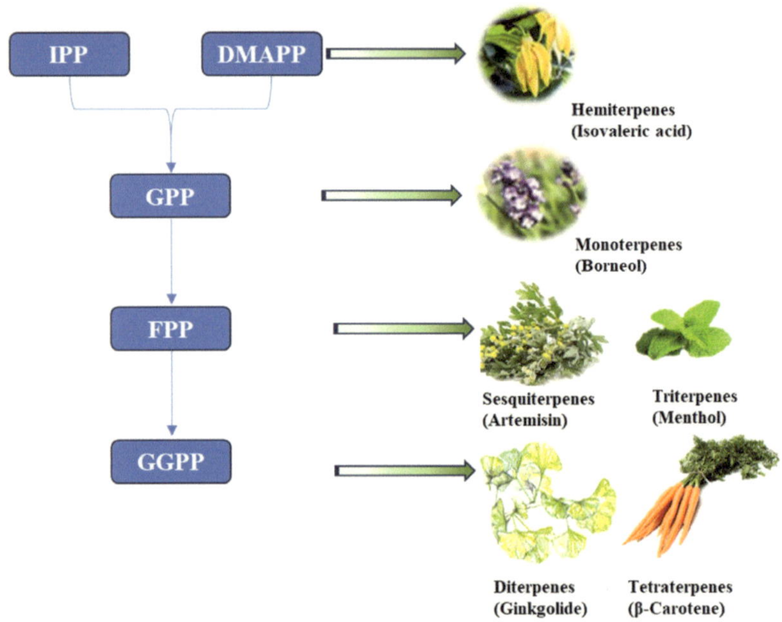

Figure 3.
Synthesis of various classes of terpenes.

4.1 Phenolics

Phenolics are a diverse group of secondary metabolites found in plants, characterized by the presence of one or more hydroxyl groups attached to a C6 aromatic ring. These compounds play crucial roles in plant defense, pigmentation, structural integrity, and interactions with the environment. Phenolic compounds include flavonoids, lignins, tannins, and phenolic acids. Their biosynthesis primarily occurs through the shikimate pathway and phenylpropanoid pathway [10].

5. Shikimate pathway (precursor formation)

The shikimate pathway provides precursors for the synthesis of phenolics. It occurs in plastids and converts simple carbohydrates into aromatic amino acids, which serve as the backbone for phenolic compounds (**Figure 4**).

6. Phenylpropanoid pathway (phenolic compound formation)

The phenylpropanoid pathway converts phenylalanine and tyrosine into a wide variety of phenolic compounds (**Figure** 5) [11].

7. Classification of phenolics

They are classified based on the number of C atoms and the structure of aromatic rings that define their structure and complexity. They are either simple phenols that

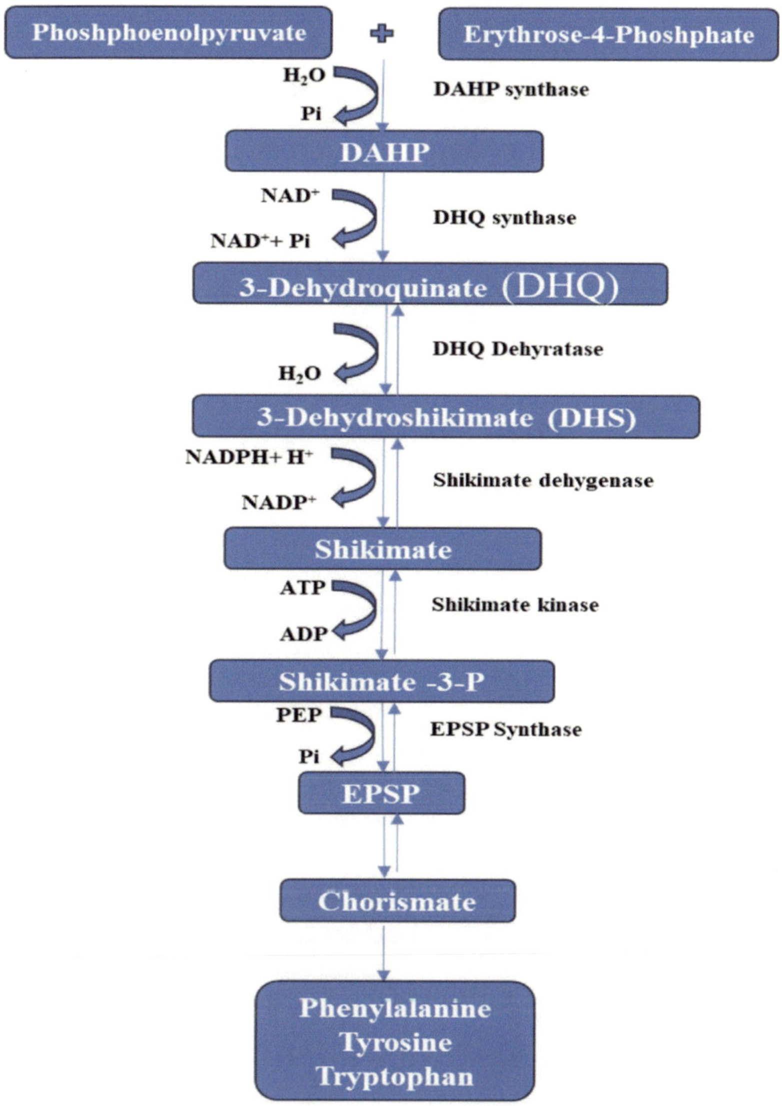

Figure 4.
Shikimate pathway: Synthesis of aromatic amino acids for the formation of phenolic compounds; DAHP (3-deoxy-d-arabino-heptulosonate-7-phosphate); EPSP (enolpyruvylshikimate 3-phosphate).

have different moieties attached to single aromatic rings or polyphenols that have two or more phenolic rings conjugated in various combinations with other moieties. Some of the phenolics and their functions are described in **Table 2**.

7.1 Alkaloids

Alkaloids are a diverse group of nitrogen-containing secondary metabolites found primarily in plants. They exhibit a wide range of biological activities, including pharmacological effects such as analgesic, antimicrobial, and anticancer properties. Alkaloids are derived from amino acids and are classified based on their precursor molecules and

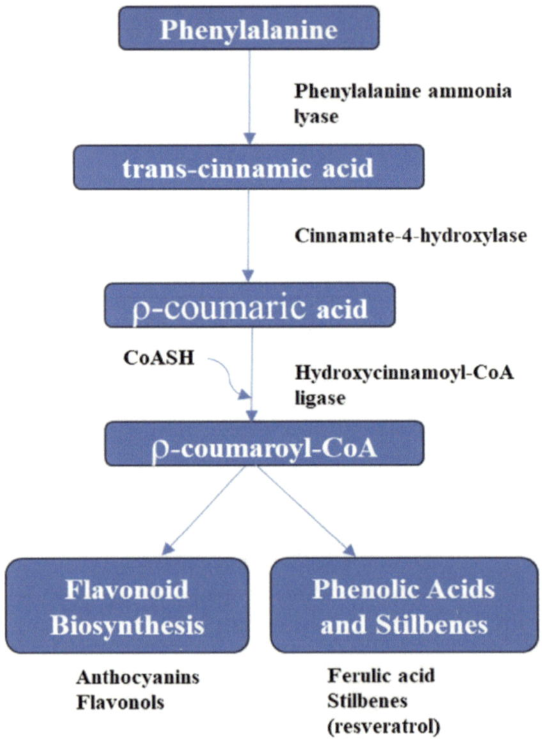

Figure 5.
Phenylpropanoid pathway: Synthesis of flavonoids and phenolic acids from aromatic amino acids.

Phenolics group	Structure	Examples and/or function
Simple phenolic acids	C6	Caffeic acid, ferulic acid, gallic acid
Flavonoids	C15	Quercetin, catechins, anthocyanins
Lignins	Polyphenols	Structural polymers providing mechanical strength
Tannins	Oligomers or polymers	Condensed and hydrolyzable tannins with antioxidant properties
Stilbenes	C6–C1–C6, C6–C2–C6	Resveratrol, piceatannol

Table 2.
Classification and function of phenolics. (Adapted from Arzani et al., [12]).

structural features [13]. Their biosynthesis involves several key pathways, including the shikimate pathway, the ornithine and lysine pathway, and the polyamine pathway [14].

8. Biosynthetic pathways of alkaloids

8.1 Shikimate pathway (aromatic amino acid-derived alkaloids)

The shikimate pathway provides precursors for alkaloids derived from tyrosine and tryptophan (**Figure 6**).

8.2 Ornithine and lysine pathway (pyrrolizidine and tropane alkaloids)

Ornithine and lysine are important precursors for the biosynthesis of pyrrolizidine and tropane alkaloids (**Figure 7**).

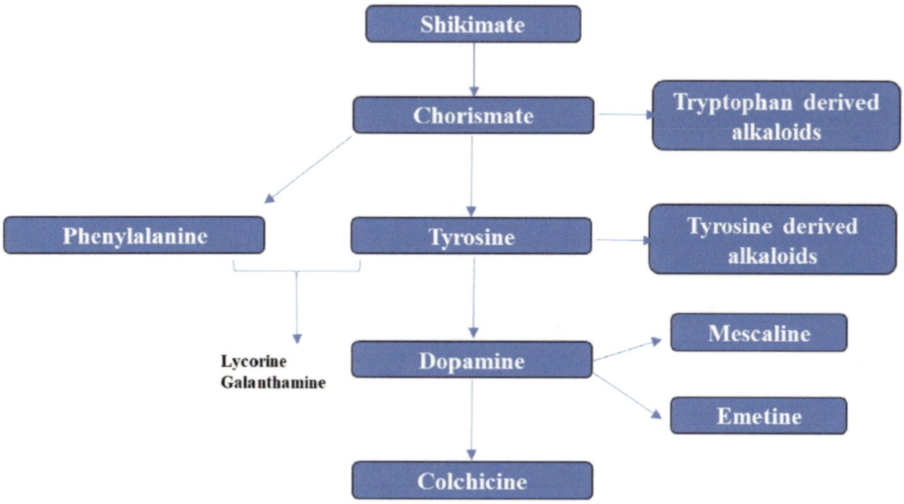

Figure 6.
Synthesis of alkaloids from aromatic amino acids.

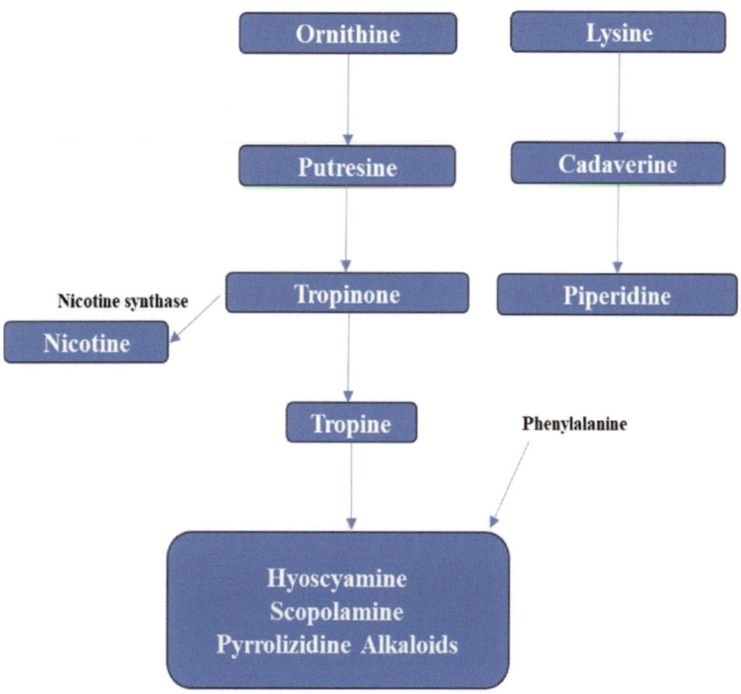

Figure 7.
Synthesis of pyrrolizidine and tropane alkaloids from ornithine and lysine.

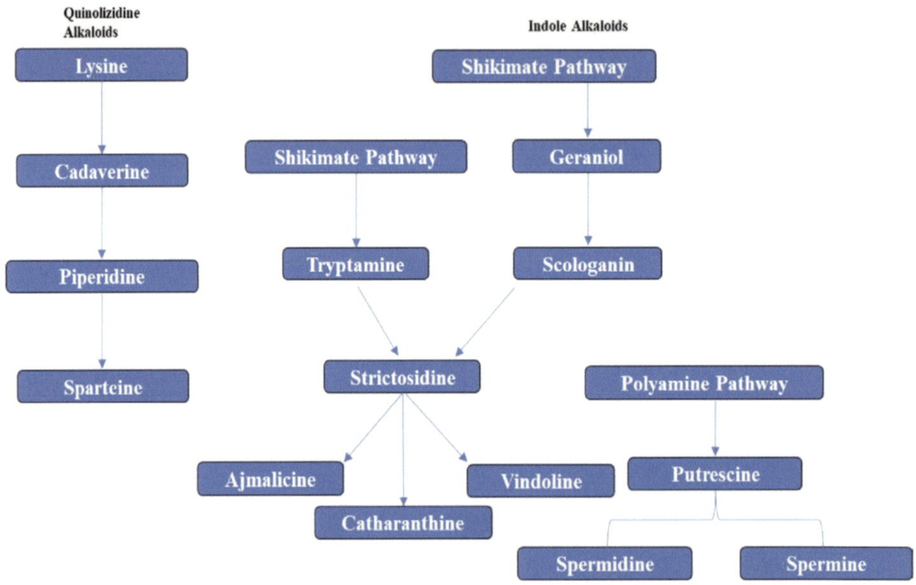

Figure 8.
Synthesis of indole and quinolizidine alkaloids.

8.3 Polyamine pathway (indole alkaloids and quinolizidine alkaloids)

The polyamine pathway involves putrescine, spermidine, and spermine, leading to the synthesis of various alkaloids (**Figure 8**).

- *Indole alkaloids*: Tryptophan is converted into strictosidine, a key precursor for alkaloids like reserpine and ajmalicine.

- *Quinolizidine alkaloids*: Lysine undergoes deamination and cyclization, leading to the formation of alkaloids such as sparteine.

9. Classification of alkaloids

Alkaloids are classified based on their biosynthetic origin as well as chemical structure. A glimpse of their classification on the basis of structure is in **Table 3**.

Category of alkaloid	Example
Indole alkaloids	Vincristine, reserpine, ajmalicine
Isoquinoline alkaloids	Morphine, berberine, papaverine
Tropane alkaloids	Atropine, scopolamine
Pyrrolizidine alkaloids	Monocrotaline, retronecine
Quinolizidine alkaloids	Sparteine, lupanine

Table 3.
Classification of alkaloids based on their structure with examples.

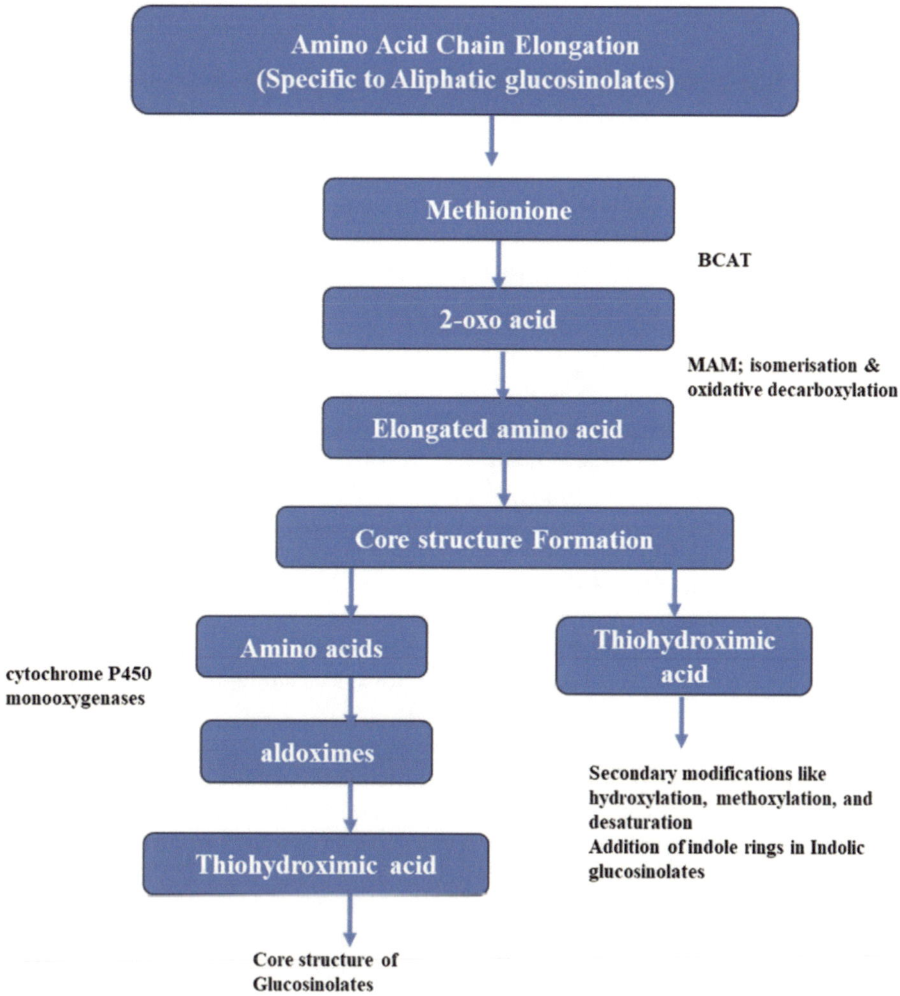

Figure 9.
Synthesis of glucosinolates.

9.1 Glucosinolates

Glucosinolates are sulfur- and nitrogen-containing secondary metabolites primarily found in the Brassicaceae family, including crops like broccoli, cabbage, mustard, and rapeseed. These compounds are well known for their role in plant defense and human health due to their hydrolysis products like isothiocyanates, which possess anticancer and antimicrobial properties [15]. Glucosinolate biosynthesis is classified into three phases: amino acid chain elongation, core structure formation, and secondary modification (**Figure 9**) [16].

10. Biosynthetic pathways of glucosinolates

- BCAT: Branched-chain amino acid transferase; MAM: methylthioalkylmalate synthase

11. Classification of glucosinolates

Based on their amino acid origin, glucosinolates are divided into:

- *Aliphatic glucosinolates*: Derived from methionine (e.g., glucoraphanin)
- *Aromatic glucosinolates*: Derived from phenylalanine (e.g., gluconasturtiin)
- *Indolic glucosinolates*: Derived from tryptophan (e.g., glucobrassicin)

11.1 Secondary metabolites in plant development

Secondary metabolites, such as glucosinolates and flavonoids, play crucial roles in regulating the growth and development of plants, as well as responding to various environmental conditions. Glucosinolates, in particular, are implicated in multiple regulatory mechanisms. For example, aliphatic 3-hydroxypropylglucosinolate has been shown to inhibit meristematic root growth in *Arabidopsis* and other plant species *via* the rapamycin signaling pathway, although the precise molecular mechanism behind this regulation remains unclear [17, 18]. In addition, aliphatic glucosinolates in *Vicia faba* and *Arabidopsis* contribute to stomatal closure through a peroxidase-mediated receptor kinase mechanism [19]. Notably, the upregulation of the myrosinase gene *TGG1* in guard cells has been associated with active control of stomatal movements through various signaling pathways [20].

Indole glucosinolates also contribute to developmental regulation. Indole-3-carbinol, an immediate breakdown product of indole glucosinolate, negatively affects root growth in *Arabidopsis* following exogenous wounding. This inhibitory effect is mediated by the binding of indole-3-carbinol to an allosteric site on a key enzyme, disrupting auxin signaling by affecting the interaction between auxin and its receptor TIR1 [21].

Flavonoids, a diverse class of phenolic compounds, are equally important in modulating plant development under both normal and stress conditions. Alterations in flavonoid biosynthesis have been shown to influence various physiological and developmental processes [22]. Flavonoids contribute to oxidative stress mitigation by scavenging reactive oxygen species (ROS) and modulating developmental processes in guard cells, leaves, and roots [23, 24]. *In vitro* studies using *Arabidopsis* have confirmed that flavonoids function as natural inhibitors of polar auxin transport [25]. This was further supported by investigations of the *transparent testa 4 (tt4)* mutant, deficient in chalcone synthase, which showed phenotypes including reduced plant height, shorter roots, and disrupted auxin transport. These effects were attributed to the accumulation of naringenin, a key precursor in the flavonoid biosynthetic pathway.

11.2 Developmental regulation of secondary metabolite production

Secondary metabolites play pivotal roles in plant defense, signaling, and ecological interactions. Their production is tightly regulated during plant development, varying with tissue type, developmental stage, environmental conditions, and genetic regulatory mechanisms. Plants possess specialized tissues and structures, such as trichomes and laticifers, where secondary metabolites are stored and expressed in response to various biotic and abiotic stresses. Developmental regulation of these metabolites is also evident, as seen with tannins, which are abundant in unripe fruits but reduce to more tolerable levels as the fruit ripens. Additionally, environmental factors like drought, heat, and light significantly influence the synthesis of secondary metabolites. Transcription factors (TFs) play a crucial role in modulating their production, aligning it with the plant's developmental stage and physiological demands. These aspects are elaborated in the following sections.

11.2.1 Tissue-specific expression

Plants can effectively localize their defense or attractants since secondary metabolite production frequently takes place in certain tissues or organs. Artemisinin in *Artemisia annua* is one of several specialized metabolites, including alkaloids, terpenoids, and flavonoids, that are distinctly produced in the outer two apical cells of glandular trichomes [26]. Its biosynthesis is tightly regulated and varies depending on the tissue type, developmental stage of the plant, and cultivar, reaching its highest levels during the flowering stage in the trichomes of floral buds. Various extrinsic factors such as light intensity, hormonal signals like jasmonic acid, and the timing of harvest significantly influence artemisinin production [27–29]. In contrast, alkaloids like nicotine are synthesized predominantly in the roots and subsequently transported to the vacuoles of aerial parts through the plant's vascular system. This process involves a highly interconnected regulatory network that governs both the biosynthesis and transport of nicotine, mediated by various transport proteins. Since nicotine production incurs a substantial metabolic cost, it is typically maintained at basal levels unless triggered by biotic stress, exposure to methyl jasmonates, or the removal of apical meristems, the latter affecting auxin levels and indirectly influencing nicotine synthesis [30]. The biosynthesis of nicotine is driven by jasmonate-induced structural genes *NICOTINE1* (NIC1) and *NICOTINE2* (NIC2), which convert precursors into nicotine [31]. To facilitate its storage and movement, transporter proteins like *NtMATE1* and *NtMATE2*, belonging to the multidrug and toxic compound extrusion (MATE) family, are co-expressed alongside these biosynthetic genes. This coordination ensures that newly synthesized nicotine is sequestered into the vacuoles of root cells, ready for translocation through the vascular system [32]. The permease protein *NUP1* further aids in nicotine uptake from the apoplast into the cells [33]. Both *MATE* transporters and *NUP1* function as positive regulators of *NIC* genes, highlighting the synchronized interaction between synthesis and transport mechanisms for efficient nicotine transfer to aerial tissues. In these tissues, the secondary transporter *Nt-JAT1*, a proton antiporter, unloads nicotine from the vascular system into vacuoles [34], while another MATE family transporter, *Nt-JAT2*, performs vacuolar sequestration of nicotine arriving *via* the xylem in leaves [35]. Similarly, the accumulation of certain flavonoids and anthocyanins in fruit skins and flower petals enhances coloration, which in turn attracts pollinators [36]. In apples, skin coloration is attributed to the accumulation of anthocyanin, a flavonoid glycoside derivative synthesized from phenylalanine. Key structural genes in this biosynthetic pathway include *phenylalanine ammonia lyase* (PAL), *chalcone synthase* (CHS), *chalcone isomerase* (CHI), *dihydroflavonol 3-hydroxylase* (F3H), *dihydroflavonol-4-reductase* (DFR), *anthocyanidin synthase* (ANS), and *UDP-glucose flavonoid 3-O-glucosyltransferase* (UFGT). The expression of these genes is regulated by multiple transcription factors from the MYB, WD40, and bHLH families. *MdMYB10* acts as a positive regulator of apple skin color, whereas a subgroup 4 R2R3-MYB, also referred to as *MdMYB10* in some contexts, functions as its repressor. This repressor binds to the promoter of *MdMYB10* and suppresses its expression, thereby reducing anthocyanin accumulation during fruit ripening [37, 38]. Moreover, DNA methylation levels at different regions of the *MdMYB10* gene modulate its expression. Methylation within the gene body in *mTotal* and *mCG* contexts positively correlates with anthocyanin accumulation, while methylation in the *mCHG* context at the upstream promoter region correlates negatively, suggesting a complex epigenetic regulation of this trait [39]. In cucumbers, mutations affecting anthocyanin biosynthesis have also been documented. Mutations in the exonic region of *CsMYB60* result in white to pale yellow fruit skin, whereas mutations in its intronic region led to light yellow coloration, indicating that introns play a regulatory role in gene expression for anthocyanin synthesis [40]. As a defense mechanism against herbivores, the seeds of many plant species accumulate phenolic compounds and glucosinolates, providing chemical protection during vulnerable developmental stages.

11.2.2 Developmental stage-dependent accumulation

The synthesis and accumulation of secondary metabolites in plants often coincide with specific developmental stages. During the seedling phase, phenolics and flavonoids accumulate in higher amounts to protect young tissues from ultraviolet (UV) radiation and potential infections. In the vegetative stage, plants typically increase the production of terpenoids and alkaloids in their leaves as a defense mechanism against herbivores. As the plant enters its reproductive phase, pigment-related metabolites such as anthocyanins accumulate in flowers and fruits to attract pollinators and aid in seed dispersal. In cucumbers, the development of brown peel color is associated with changes in flavonoid accumulation across different fruit developmental stages, from growth to maturation. A combined flavonoid-targeted metabolomic and transcriptomic analysis revealed that at 12 days post-pollination (DPP), flavonoid levels in the peel began to rise sharply compared to earlier growth stages. This increase was accompanied by a notable upregulation of key genes involved in the early steps of the flavonoid biosynthesis pathway, including *chalcone synthase* (CHS), *chalcone isomerase* (CHI), and *dihydroflavonol 3-hydroxylase* (F3H). Interestingly, the brown coloration in cucumber peel is primarily attributed to the accumulation of naringenin chalcone, a yellowish pigment. Under normal conditions, this compound is rapidly converted into naringenin by the enzyme encoded by *CHI*. However, in the brown-colored peel samples, the expression of *CHI* genes was significantly downregulated, inhibiting this conversion. As a result, naringenin chalcone accumulated, imparting a characteristic brown color to the cucumber peel [41]. Additionally, during the senescence phase of plant development, levels of flavonols and other antioxidant compounds increase to help combat oxidative stress and minimize cellular damage associated with aging.

11.2.3 Environmental influences

Environmental factors have a significant impact on secondary metabolite profile and quantity. For instance, flavonoid production (such as that of anthocyanins in *Arabidopsis*) is enhanced by UV-B and blue light. Flavonoid production in grape skin is reported to be under the influence of both light and temperature. Sufficient anthocyanin accumulation occurs at low temperatures (15°C) complemented with light treatment, but dark treatment or a high temperature (35°C) suppresses it [42]. Deliberate induction that promoted the development of red peel of mango skin color under artificial light in postharvest storage when inflicted on the unripe green ones indicates the role of light in inducing the flavonoid biosynthesis pathway. Expression study of key flavonoid biosynthetic genes, including *MiF3H*, *MiFLS*, *MiLAR*, *MiANS*, *MiUFGT1*, and *MiUFGT3,* and their positive regulators, MiMYB22, MiMYB12,

MiHY5, and *MiHYH*, showed upregulation in response to light [43]. Likewise, light regulates artemisinin production by simultaneously activating positive regulators and repressing negative regulators of its biosynthetic pathway. Specifically, light-induced activation of genes such as *amorpha-4,11-diene synthase*, *cytochrome P450 monooxygenase*, *cytochrome P450 reductase*, *aldehyde dehydrogenase 1*, *artemisinic aldehyde*, and *glandular secretory trichome (GST)* enhances artemisinin accumulation. This is further supported by the light-induced repression of negative regulators like *AaMYB15*, which acts as a suppressor in this pathway [29]. Additionally, methyl jasmonate promotes the expression of *AaWRKY9*, a transcription factor that directly binds to the promoters of positive regulator genes such as *AaDBR2* and *AaGSW1*, thereby augmenting artemisinin production [28]. Heat can increase alkaloid biosynthesis in certain species, such as CrMAPK3 in *Catharanthus roseus*, causing stress-induced accumulation of monoterpenoid indole alkaloids [44]. Cold temperature stress of 15°C to watermelon caused accumulation of phenolics by activating their biosynthesis as well as inhibiting their oxidation [45]. However, prolonged cold exposure to grape leaves caused a considerable decrease in phenolics—caffeic acid, *p*-coumaric acid, ferulic acid, and a caffeic acid derivative—and also decreased antioxidant capacity [46]. Carbon-based metabolites such as terpenoids and phenolics can rise when nitrogen levels are low. Contrastingly, alkaloids such as vinblastine, vincristine, and catharanthine accumulate at higher levels in seedlings when nitrogen is supplied to UV-B-irradiated plants, probably as a physiological response to alleviate the damage caused by UV-B and free radicals generated as its result [47]. Phytoalexins, glucosinolates, or alkaloids are produced by biotic stressors such as insect herbivory or pathogen attack *via* the signaling pathways of jasmonate and salicylate.

11.2.4 Role of transcription factors and signaling pathways

Secondary metabolite pathways are mostly controlled by transcriptional regulation. Flavonoid and anthocyanin biosynthesis genes, such as PAP1 in Arabidopsis, are regulated by the MYB, bHLH, and WD40 complexes [48]. Defense-related secondary metabolism, including the synthesis of camalexin, is mediated by WRKY (conserved WRKYGQK sequence) transcription factors [49]. Terpenoid indole alkaloid production is modulated by AP2/ERF factors (e.g., ORCA3 in *Catharanthus roseus*). Plant hormones, specifically ethylene, salicylic acid, and jasmonic acid (JA), are important signaling molecules that activate secondary metabolic genes. Secondary metabolite gene clusters are similarly influenced by histone changes and DNA methylation.

12. Mechanisms preventing autotoxicity of secondary metabolites in plants

Efficient transport and storage are essential to prevent autotoxicity and to ensure metabolite availability at functional sites. Many phenolics, alkaloids, and anthocyanins are sequestered into vacuoles *via* ATP-binding cassette (ABC) transporters and multidrug and toxic compound extrusion (MATE) transporters (**Figure 10**). For example, anthocyanins in grape skins are stored in vacuoles for pigmentation and antioxidant functions. Glandular trichomes act as specialized secretory structures for the synthesis and storage of terpenoids and flavonoids, creating reservoirs on plant surfaces. For instance, essential oils are stored in trichomes of *Mentha* and *Lavandula* species [50]. Some specialized cells called laticifers are found in plants like *Papaver*

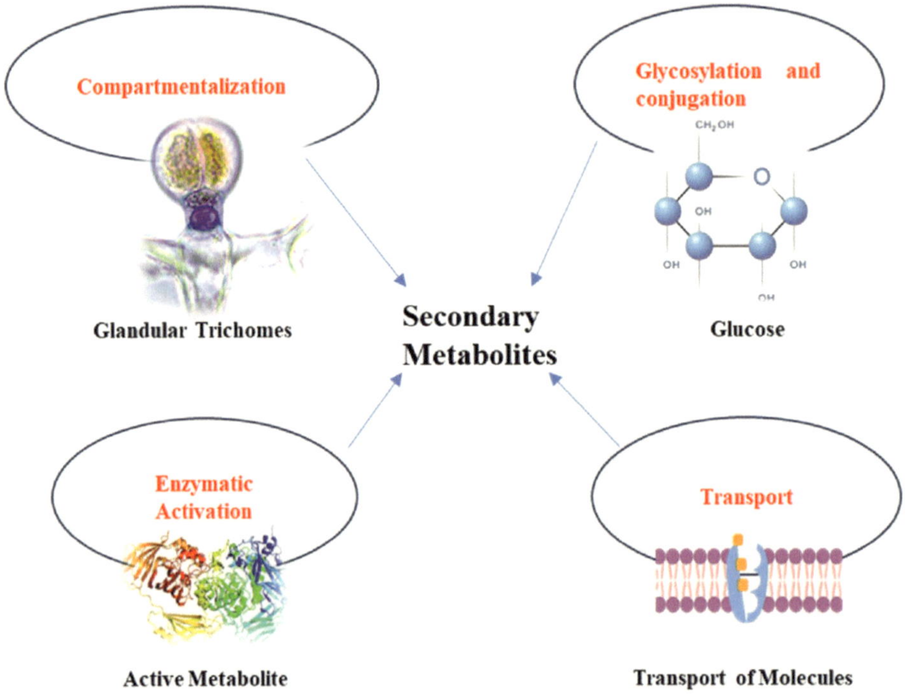

Figure 10.
Strategies to prevent autotoxicity of secondary metabolites in plants. Plants use vacuolar storage, glycosylation, transporter efflux, and tissue-specific expression to avoid self-toxicity.

somniferum; these cells store latex rich in alkaloids such as morphine, serving both defense and medicinal functions. Similarly, rubber is stored in the laticifers of *Hevea brasiliensis*.

13. Secondary metabolites in plant defense

Plants have evolved a diverse array of secondary metabolites (SM) to defend themselves against herbivores, pathogens, and environmental stresses. Over millions of years, natural selection has driven the development of these specialized compounds, allowing plants to survive in hostile environments. Early land plants, lacking physical mobility, relied on chemical defenses to deter herbivores and inhibit pathogen growth. Secondary metabolites such as alkaloids, phenolics, and terpenoids emerged as crucial defense molecules. Alkaloids (e.g., caffeine, nicotine) evolved as neurotoxins that disrupt the nervous systems of herbivores, reducing feeding activity. Phenolics (e.g., tannins, flavonoids) were developed to deter herbivory through astringency and antimicrobial properties. Terpenoids (e.g., limonoids, phytoalexins) functioned as toxic deterrents and signaling molecules, helping plants resist insect attacks and microbial infections [51]. The evolution of inducible defenses allowed plants to produce secondary metabolites only when threatened, conserving energy while ensuring survival. For example, jasmonic acid and salicylic acid pathways evolved to regulate secondary metabolite production in response to herbivore and pathogen attacks [52]. Additionally, plants developed volatile organic compounds (VOCs) that serve as airborne signals to attract natural predators of herbivores or alert neighboring plants of impending danger. Coevolution with herbivores has further shaped plant defense strategies. Some herbivores adapted to detoxify plant metabolites, prompting plants to evolve more complex and potent compounds. This evolutionary arms race continues, demonstrating the dynamic role of secondary metabolites in plant survival and ecological interactions. The interplay between genes, their expression, and the resulting traits is genuinely fascinating. For example, the manipulation of secondary metabolites not only boosts plant resilience but also holds immense potential for innovations in agriculture and medicine.

One exciting area is how plants produce unique secondary metabolites like flavonoids, alkaloids, and terpenoids, which have enormous benefits for plant resilience and human health. For instance, flavonoids not only contribute to the plant's UV filtration, nitrogen fixation, and floral pigmentation but also act as powerful antioxidants when consumed by humans. This plays a vital role in reducing the risk of chronic diseases. On the other hand, alkaloids can serve as defense compounds for plants, deterring herbivores and pathogens, but in controlled doses, they also offer medicinal properties. Meanwhile, terpenoids, the largest class of secondary metabolites, are involved in a myriad of processes, from growth regulation to attracting pollinators with their aromatic properties. Alkaloids, such as nicotine and morphine, act as toxic deterrents against herbivores and pathogens by interfering with their nervous systems or metabolic pathways. Phenolics, including flavonoids, tannins, and lignin, provide structural reinforcement and antioxidant properties, preventing microbial infections and deterring herbivory. Terpenoids, such as limonoids and phytoalexins, contribute to direct toxicity against pathogens, insect repellence, and allelopathy, which suppresses competing plant growth. Plants also employ secondary metabolites as signaling molecules in plant-microbe interactions, triggering defense pathways upon pathogen attack. The production of phytoalexins, for instance, is rapidly induced in response to microbial invasion, preventing disease progression. Additionally, volatile organic compounds (VOCs) serve as airborne signals, attracting natural predators of herbivores or warning neighboring plants of impending danger [53]. The biotechnological potential of secondary metabolites is vast, as they can be harnessed for developing biopesticides, pharmaceuticals, and stress-resistant crop varieties. Understanding their biosynthesis and regulatory pathways is crucial for enhancing plant defense mechanisms through genetic engineering and breeding strategies.

13.1 Defense against herbivores

Plants have evolved intricate defense strategies to protect themselves from herbivorous attacks, broadly categorized into direct and indirect defenses, largely mediated by secondary metabolites. Direct defenses involve compounds that deter or harm herbivores physically or chemically. For instance, toxins such as alkaloids (e.g., nicotine from *Nicotiana tabacum* and morphine from *Papaver somniferum*) can disrupt herbivore nervous systems, while cyanogenic glycosides release hydrogen cyanide upon tissue damage, causing acute toxicity. Antifeedants, including terpenoids and tannins, make plant tissues less palatable or digestible—tannins, in particular, bind to dietary proteins and digestive enzymes, inhibiting their function. Additionally, digestive enzyme inhibitors like protease and amylase inhibitors interfere with herbivore digestion, reducing nutrient absorption and growth. On the other hand, indirect defenses function by recruiting natural enemies of herbivores [54]. Volatile organic compounds (VOCs), such as methyl jasmonate, are released when plants are attacked and act as airborne cues to attract predators and parasitoids—for example, maize (*Zea mays*) emits VOCs to lure parasitic wasps. Extrafloral nectaries, found in species like *Vicia faba*, secrete nectar to attract protective insects like ants, which aggressively fend off herbivores. Moreover, plants exhibit induced defenses, where herbivory triggers systemic biochemical responses, often mediated by signaling molecules like jasmonic acid. This leads to the enhanced production of defense-related secondary metabolites such as glucosinolates, phenolics, and terpenoids, strengthening the plant's resistance to further attack. These dynamic defense systems illustrate how plants employ both chemical warfare and ecological partnerships to counteract herbivore threats effectively [55, 56].

13.2 Defense against pathogens

Plants possess a sophisticated immune system to combat pathogenic attacks, in which secondary metabolites and associated proteins play central roles. One of the primary defense mechanisms involves the synthesis of phytoalexins, which are low-molecular-weight antimicrobial compounds produced de novo at infection sites in response to pathogen invasion. These compounds are not stored but synthesized upon detection of a threat and exhibit strong antifungal and antibacterial activities. For instance, isoflavonoids act as key phytoalexins in legumes, while terpenoids serve this role in cereals such as rice and maize. Another vital defense strategy involves the activation of pathogenesis-related (PR) proteins, which are often co-induced with secondary metabolites. These include chitinases, which break down fungal cell walls composed of chitin; glucanases, which degrade β-glucans in fungal structures; and proteinase inhibitors, which disrupt proteolytic enzymes in pathogens. Together, these proteins limit pathogen growth and spread within plant tissues.

Beyond localized responses, plants also mount systemic defenses such as Systemic Acquired Resistance (SAR) and Induced Systemic Resistance (ISR). SAR is typically activated by salicylic acid (SA) and provides long-lasting immunity throughout the plant against biotrophic pathogens. In contrast, ISR is triggered by jasmonic acid (JA) and ethylene, offering resistance against necrotrophic pathogens and some insect pests. Both forms of systemic resistance involve transcriptional reprogramming and the accumulation of defense-related metabolites and proteins in uninfected tissues, enhancing overall plant immunity. Additionally, many secondary metabolites, including saponins, phenols, and flavonoids, exhibit direct antimicrobial properties. These compounds interfere with pathogen viability by disrupting their cell membranes, altering permeability, or inhibiting vital metabolic processes. Through this multilayered chemical and protein-based defense arsenal, plants effectively mitigate pathogen infections and preserve their health.

13.3 Defense against other stresses

Plants have evolved a diverse array of secondary metabolites that serve crucial protective roles under various abiotic and biotic stress conditions. One important function of secondary metabolites is UV protection. Flavonoids, particularly those accumulated in the epidermal tissues of leaves, act as natural sunscreens by absorbing harmful UV-B radiation, thereby shielding cellular DNA from damage and reducing the risk of mutagenesis [57]. These compounds help maintain cellular homeostasis under intense solar radiation, especially in high-altitude or open-field environments.

Secondary metabolites also play a vital role in oxidative stress protection. Under environmental stresses such as drought, salinity, or high light intensity, plants generate reactive oxygen species (ROS), which can damage proteins, lipids, and nucleic acids. Antioxidant secondary metabolites like flavonoids, carotenoids, and phenolic acids neutralize these ROS, thereby mitigating oxidative damage and preserving cellular integrity [58]. For instance, flavonoids are known to donate electrons to ROS, stabilizing them and preventing chain reactions that could damage cell membranes.

Another fascinating ecological role of secondary metabolites is allelopathy. In this phenomenon, certain plants release bioactive compounds into the soil through root exudation, leaf leaching, or decomposition of plant litter, which inhibit the germination, growth, or nutrient uptake of neighboring competing plant species. A well-documented example is juglone, a naphthoquinone released by black walnut (*Juglans nigra*), which suppresses the growth of surrounding vegetation by disrupting cellular respiration and root development (**Figure 11**) [59]. Allelopathy is not only an effective strategy for reducing competition but also influences plant community dynamics and agricultural crop planning.

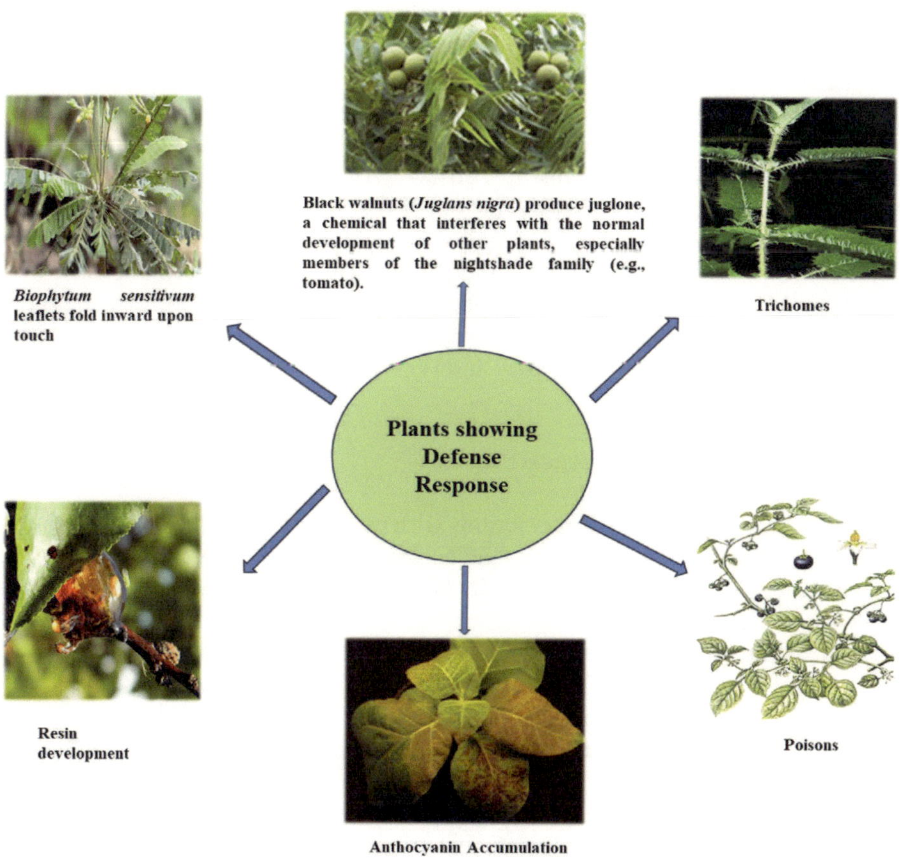

Figure 11.
Plants exhibiting a defense response.

14. Signal transduction in defense responses

The plant immune system is a complex network of signaling pathways, with jasmonic acid (JA), salicylic acid (SA), and ethylene playing pivotal roles in coordinating defense responses against various biotic stresses, including herbivory and pathogen attacks. Jasmonic acid (JA) is a crucial hormone that regulates plant defense against herbivores and necrotrophic pathogens (those that feed on dead or decaying tissue) by triggering the production of secondary metabolites such as terpenoids and alkaloids, which have toxic effects on herbivores and inhibit pathogen growth [60]. JA also activates genes involved in plant resistance, such as those encoding proteinase inhibitors that impair herbivore digestion.

On the other hand, salicylic acid (SA) is essential in defending against biotrophic pathogens, which require living host tissue for survival. SA plays a vital role in the activation of Systemic Acquired Resistance (SAR), a form of long-lasting immunity that occurs after the initial pathogen attack [61]. SA enhances the expression of pathogenesis-related (PR) proteins, which directly inhibit pathogen growth and aid in the establishment of systemic immunity. Ethylene, a gaseous plant hormone, interacts with both the JA and SA pathways, modulating defense gene expression. Ethylene is involved in various stress responses, including resistance to pathogens and the regulation of fruit ripening [62]. It enhances the activation of certain defense genes, especially those involved in the production of antimicrobial compounds like phytoalexins, while also promoting the expression of jasmonate-regulated genes during herbivore attacks.

The interactions between JA, SA, and ethylene are not merely additive but also involve cross-talk that fine-tunes defense responses. These pathways often exhibit antagonistic or synergistic relationships, ensuring that the plant's defense mechanisms are appropriately tailored to the nature of the stress. For instance, while JA primarily activates defense responses against herbivores and necrotrophs, SA is generally more involved in defense against biotrophic pathogens. The balance between these pathways determines whether the plant responds more effectively to one type of stress over another [63]. This intricate cross-talk ensures that plants can prioritize defense responses based on the severity and type of threat encountered.

15. Ecological roles and interactions

Plants engage in a variety of interactions with other organisms, which contribute significantly to their survival, growth, and ecological balance. These interactions can be broadly categorized into plant–animal, plant-microbe, and plant–plant interactions.

15.1 Plant–animal interactions

Plants depend on animals for various ecological functions. One prominent example is pollinator attraction, where plants use floral pigments and scents to attract pollinators like bees, butterflies, and birds. Nectar secondary metabolites act as a plant interface for a complex interaction with insects and other organisms [64]. The bright colors of flowers and the aromatic volatile compounds they release are specifically designed to attract these pollinators, facilitating cross-pollination and ensuring genetic diversity within plant populations [65]. Additionally, as plants are sessile in nature, they often depend on animals to move their seeds away from the parent plant, or seed dispersal, thus minimizing competition and promoting the spread of the species. For instance, fruits, seeds, and nuts containing secondary metabolites containing either edible or sticky substances attract animals that transport the seeds to new locations where they can germinate [56]. It has been reported that 52% of the seed-producing plants are dispersed by the animals [66].

15.2 Plant–microbe interactions

Secondary metabolites from plants are shown to affect the type and composition of the microbiome around the roots [67, 68]. Plants also interact with a wide range of microbes, from symbiotic relationships to antagonistic ones. One of the most well-known symbiotic relationships is nitrogen fixation, where plants, particularly legumes, host nitrogen-fixing bacteria such as *Rhizobium* in specialized root nodules. These bacteria convert atmospheric nitrogen into forms that plants can use, enhancing soil fertility and reducing the need for synthetic fertilizers [69]. On the other hand, plants can also experience antagonistic interactions with microbes. Certain fungi, bacteria, and viruses can act as pathogens, infecting plants and causing diseases. In response, plants produce a variety of defense mechanisms, including the synthesis of antimicrobial compounds (phytoalexins) and the activation of immune signaling pathways [70]. Glucosinolates in *Brassica* sp. affect soil microbiota to control diseases. It is metabolized by myrosinase into toxic and deterrent isothiocyanates, nitriles, or other products, which are used to deter herbivores [71]. Coumarins secreted in plant roots have been reported to increase the bioavailability of iron through reduction and chelation [72] and microbial function [73]. Triterpenes in *Arabidopsis* have been reported to have a role in plant-microbe interaction [74].

15.3 Plant–plant interactions (allelopathy)

Plants can also interact with each other in ways that influence community structure and biodiversity. Allelopathy refers to the chemical interactions between plants, where one plant releases secondary metabolites into the environment that inhibit the germination, growth, or establishment of nearby plants [75]. These chemicals, such as juglone produced by walnut trees, help plants secure space, resources, and light by reducing competition from neighboring species [76]. While allelopathy can provide a competitive advantage to the plant producing the chemicals, it can also have negative effects on biodiversity by suppressing other plant species. The secondary metabolites released during allelopathic interaction are autotoxic, as reported in seaweed [77]. The allelopathic interaction is chemically mediated, and terpenoids and strigolactones are believed to be involved in the allelopathic interaction between root-soil interactions [78].

The interactions between plants, animals, microbes, and even between plant roots themselves create a complex web that shapes a plant's fitness, survival, performance, species diversity, resource utilization, and chemical communication. Secondary metabolites play a pivotal role in these interactions, emphasizing the need to study the underlying chemical signals and molecular mechanisms that govern them.

16. Conclusions and future prospects

Secondary metabolites are the unsung heroes of the plant kingdom. Although produced in small quantities, they play a vital role in ensuring plant survival under changing climatic conditions. These bioactive compounds act as a protective shield, enabling plants to withstand and adapt to diverse growth environments and stress factors. Both biotic and abiotic stresses activate signal transduction pathways that lead to the synthesis of secondary metabolites, which, in turn, strengthen the plant's natural defense mechanisms. Interestingly, secondary metabolites serve a dual purpose, i.e., they function both as regulatory intermediates of primary bioactive compounds and as end products involved in essential physiological processes. Despite their significance, many of the molecular links and mechanisms underlying these defense responses remain unexplored. The characterization of metabolic networks and pathways associated with these responses represents a fascinating yet under-investigated area of plant biology.

Technological advancements in omics sciences have greatly enhanced our understanding of the regulatory networks and complex metabolic pathways involved in the biosynthesis of secondary metabolites. The future of secondary metabolite research lies in integrating artificial intelligence (AI) with genomics, transcriptomics, proteomics, metabolomics, and interactomics to unravel their roles in ecological interactions, regulatory mechanisms, and biosynthetic processes. An equally promising area is the study of plant-microbiome interactions mediated by these secondary metabolites, which hold the potential to improve crop yields and develop climate-resilient varieties. AI and machine learning can also be employed to predict novel metabolites and identify strategies to optimize their biosynthesis. Together, these approaches can harness the vast potential of secondary metabolites across agriculture, medicine, and industry, driving sustainable development and bio-innovation. Unlocking the full potential of these bioactive compounds could ultimately shape a more resilient, productive, and sustainable future for both humanity and the natural environment.

Author details

Archana Watts[1*], Yuvonne Angel Lyngdoh[2], Anshul Watts[3] and Shruti Sinha[4]

1 Division of Plant Physiology, ICAR-IARI, New Delhi, India

2 ICAR-Central Potato Research Station, Shillong, Meghalaya, India

3 ICAR-National Institute of Plant Biotechnology, New Delhi, India

4 Division of Genomic Resources, ICAR-NBPGR, New Delhi, India

*Address all correspondence to: arch.9863@gmail.com

References

[1] Wink M. Plant secondary metabolism: Diversity, function and its evolution. Natural Product Communications. 2008;**3**:1205-1216

[2] Leroux H. Discovery of salicine. Journal of Medicinal Chemistry. 1830;**6**:340-432

[3] Fisch C. William withering: an account of foxglove and some of its medical uses. 1785-1985. Journal of the American College of Cardiology. 1985;**5**(5):1A-2A. DOI: 10.1016/S0735-1097(85)80456-3

[4] Schmitz R. Friedrich Wilhelm Serturner and the discovery of morphine. Pharmacy in History. 1985;**27**:61-74

[5] Achan J, Talisuna AO, Erhart A, Yeka A, Tibenderana JK, Baliraine FN, et al. Quinine, an old anti-malarial drug in a modern world: role in the treatment of malaria. Malaria Journal. 2011;**10**(1):144. DOI: 10.1186/1475-2875-10-144

[6] Burrell RC. Phytochemistry. What it is and how it has developed? Journal of Chemical Education. 1937;**14**(11):520. DOI: 10.1021/ed014p520

[7] Dewick PM. Medicinal Natural Products: A Biosynthetic Approach. HobokenNJ, USA: John Wiley & Sons; 2002

[8] Kirby J, Keasling JD. Metabolic Engineering for Terpenoid Production in Microbial Hosts. Nature Reviews. Microbiology. 2009;**7**(12):970-980

[9] Vranová E et al. Network Analysis of the MVA and MEP Pathways for Isoprenoid Synthesis. Trends in Plant Science. 2013;**18**(5):260-268

[10] Cheynier V, Comte G, Davies KM, Lattanzio V, Martens S. Plant phenolics: Recent advances on their biosynthesis, genetics, and ecophysiology. Plant Physiology and Biochemistry. 2013;**72**:1-20. DOI: 10.1016/j.plaphy.2013.05.009

[11] Fraser CM, Chapple C. The Phenylpropanoid Pathway in Arabidopsis. The Arabidopsis Book. 2011;**9**:e0152

[12] Arzani V, Soleimani M, Fritsch T, Jacob UM, Calabrese V, Arzani A. Plant polyphenols, terpenes, and terpenoids in oral health. Open Medicine. 2025;**20**(1):20251183

[13] Facchini PJ. Alkaloid Biosynthesis in Plants: Biochemistry, Cell Biology, Molecular Regulation. Annual Review of Plant Physiology and Plant Molecular Biology. 2001;**52**:29-66

[14] Kutchan TM. Alkaloid Biosynthesis: The Basis for Metabolic Engineering of Medicinal Plants. The Plant Cell. 1995;**7**(7):1059-1070

[15] Halkier BA, Gershenzon J. Biology and Biochemistry of Glucosinolates. Annual Review of Plant Biology. 2006;**57**:303-333

[16] Sønderby IE, Geu-Flores F, Halkier BA. Biosynthesis of glucosinolates - gene discovery and beyond. Trends in Plant Science. 2010;**15**:283-290. DOI: 10.1016/j.tplants.2010.02.005

[17] Kerwin RE, Jimenez-Gomez JM, Fulop D, Harmer SL, Maloof JN, Kliebensteina DJ. Network quantitative trait loci mapping of circadian clock outputs identifies metabolic pathway-to-clock linkages in Arabidopsis. The Plant Cell. 2011;**23**:471-485. DOI: 10.1105/tpc.110.082065

[18] Malinovsky FG, Thomsen MLF, Nintemann SJ, Jagd LM, Bourgine B, Burow M, et al. An evolutionarily young defense metabolite influences the root growth of plants via the ancient TOR signaling pathway. eLife. 2017;**6**:e29353

[19] Hossain MS, Ye W, Hossain MA, Okuma E, Uraji M, Nakamura Y, et al. Glucosinolate degradation products, isothiocyanates, nitriles, and thiocyanates, induce stomatal closure accompanied by peroxidase-mediated reactive oxygen species production in Arabidopsis thaliana. Bioscience, Biotechnology, and Biochemistry. 2013;**77**:977-983

[20] Bhat R, Vyas D. Myrosinase: Insights on structural, catalytic, regulatory, and environmental interactions. Critical Reviews in Biotechnology. 2019;**39**:508-523

[21] Katz E, Nisani S, Yadav BS, Woldemariam MG, Shai B, Obolski U, et al. The glucosinolate breakdown product indole-3-carbinol acts as an auxin antagonist in roots of Arabidopsis thaliana. The Plant Journal. 2015;**82**:547-555

[22] Gayomba SR, Watkins JM, Muday GK. Flavonols regulate plant growth and development through regulation of auxin transport and cellular redox status. Recent Advances in Polyphenol Research. 2016;**5**:143-170

[23] Buer CS, Djordjevic MA. Architectural phenotypes in the transparent testa mutants of Arabidopsis thaliana. Journal of Experimental Botany. 2009;**60**:751-763

[24] Watkins JM, Chapman JM, Muday GK. Abscisic acid-induced reactive oxygen species are modulated by flavonols to control stomata aperture. Plant Physiology. 2017;**175**:1807-1825

[25] Daryanavard H, Postiglione AE, Mühlemann JK, Muday GK. Flavonols modulate plant development, signaling, and stress responses. Current Opinion in Plant Biology. 2023;**72**:102350

[26] Olsson ME, Olofsson LM, Lindahl AL, Lundgren A, Brodelius M, Brodelius PE. Localization of enzymes of artemisinin biosynthesis to the apical cells of glandular secretory trichomes of Artemisia annua L. Phytochemistry. 2009;**70**(9):1123-1128

[27] Weathers PJ, Arsenault PR, Covello PS, McMickle A, Teoh KH, Reed DW. Artemisinin production in Artemisia annua: studies in planta and results of a novel delivery method for treating malaria and other neglected diseases. Phytochemistry Reviews. 2011;**10**:173-183

[28] Fu X, Peng B, Hassani D, Xie L, Liu H, Li Y, et al. AaWRKY9 contributes to light- and jasmonate-mediated regulation of artemisinin biosynthesis in Artemisia annua. The New Phytologist. 2021;**231**(5):1858-1874

[29] Liu Y, Singh SK, Pattanaik S, Wang H, Yuan L. Light regulation of the biosynthesis of phenolics, terpenoids, and alkaloids in plants. Communications Biology. 2023;**6**(1):1055

[30] Shi Q, Li C, Zhang F. Nicotine synthesis in Nicotiana tabacum L. induced by mechanical wounding is regulated by auxin. Journal of Experimental Botany. 2006;**57**(11):2899-2907

[31] Shoji T, Hashimoto T. Smoking out the masters: transcriptional regulators for nicotine biosynthesis in tobacco. Plant Biotechnology. 2013;**30**(3):217-224

[32] Shoji T, Inai K, Yazaki Y, Sato Y, Takase H, Shitan N, et al. Multidrug

and toxic compound extrusion-type transporters implicated in vacuolar sequestration of nicotine in tobacco roots. Plant Physiology. 2009;**149**(2):708-718

[33] Hildreth SB, Gehman EA, Yang H, Lu RH, KC R, Harich KC, et al. Tobacco nicotine uptake permease (NUP1) affects alkaloid metabolism. Proceedings of the National Academy of Sciences of the United States of America. 2011;**108**(44):18179-18184

[34] Morita M, Shitan N, Sawada K, Van Montagu MC, Inzé D, Rischer H, et al. Vacuolar transport of nicotine is mediated by a multidrug and toxic compound extrusion (MATE) transporter in Nicotiana tabacum. Proceedings of the National Academy of Sciences of the United States of America. 2009;**106**(7):2447-2452

[35] Shitan N, Minami S, Morita M, Hayashida M, Ito S, Takanashi K, et al. Involvement of the leaf-specific multidrug and toxic compound extrusion (MATE) transporter Nt JAT2 in vacuolar sequestration of nicotine in Nicotiana tabacum. PLoS One. 2014;**9**(9):e108789

[36] Narbona E, del Valle JC, Arista M, Buide ML, Ortiz PL. Major flower pigments originate different colour signals to pollinators. Frontiers in Ecology and Evolution. 2021;**9**:743850

[37] Chagné D, Lin-Wang K, Espley RV, Volz RK, How NM, Rouse S, et al. An ancient duplication of apple MYB transcription factors is responsible for novel red fruit-flesh phenotypes. Plant Physiology. 2013;**161**(1):225-239

[38] Ding T, Zhang R, Zhang H, Zhou Z, Liu C, Wu M, et al. Identification of gene co-expression networks and key genes regulating flavonoid accumulation in apple (Malus × domestica) fruit skin. Plant Science. 2021;**304**:110747

[39] Li WF, Ning GX, Mao J, Guo ZG, Zhou Q, Chen BH. Whole-genome DNA methylation patterns and complex associations with gene expression associated with anthocyanin biosynthesis in apple fruit skin. Planta. 2019;**250**:1833-1847

[40] Xu X, Zhu Y, Yuan Y, Sohail H, He S, Ye Y, et al. R2R3-MYB transcription factor CsMYB60 controls mature fruit skin color by regulating flavonoid accumulation in cucumber. The Plant Journal. 2024;**119**(2):796-813

[41] He S, Ye Y, Yuan Y, Lv M, Wang M, Xu Q, et al. Insights into flavonoid biosynthesis during cucumber fruit peel coloration based on metabolite profiling and transcriptome analyses. Horticultural Plant Journal. 2023;**9**(4):763-776

[42] Azuma A, Yakushiji H, Koshita Y, Kobayashi S. Flavonoid biosynthesis-related genes in grape skin are differentially regulated by temperature and light conditions. Planta. 2012;**236**:1067-1080

[43] Zhu W, Wu H, Yang C, Shi B, Zheng B, Ma X, et al. Postharvest light-induced flavonoids accumulation in mango (Mangifera indica L.) peel is associated with the up-regulation of flavonoids-related and light signal pathway genes. Front. Plant Science. 2023;**14**:1136281

[44] Raina S, Wankhede D, Sinha A. Catharanthus roseus mitogen-activated protein kinase 3 confers UV and heat tolerance to Saccharomyces cerevisiae. Plant Signaling & Behavior. 2013;**8**(1):e22716

[45] Rivero RM, Ruiz JM, García PC, Lopez-Lefebre LR, Sánchez E, Romero L. Resistance to cold and heat stress: accumulation of phenolic compounds

in tomato and watermelon plants. Plant Science. 2001;**160**(2):315-321

[46] Król A, Amarowicz R, Weidner S. The effects of cold stress on the phenolic compounds and antioxidant capacity of grapevine (Vitis vinifera L.) leaves. Journal of Plant Physiology. 2015;**189**:97-104

[47] Guo XR, Chang BW, Zu YG, Tang ZH. The impacts of increased nitrate supply on Catharanthus roseus growth and alkaloid accumulations under ultraviolet-B stress. Journal of Plant Interactions. 2014;**9**(1):640-646

[48] Zhang H et al. Regulatory network of the Arabidopsis MYB transcription factors involved in secondary metabolism. Molecular Plant. 2011;**4**(5):761-774

[49] Chezem WR, Clay NK. Regulation of plant secondary metabolism by WRKY transcription factors. The Plant Cell. 2016;**28**(5):933-934

[50] Lange BM, Turner GW. Glandular trichomes – sites of biosynthesis, secretion and storage of plant secondary metabolites. Plant Biology. 2013;**15**(4):614-628

[51] Tholl D. Biosynthesis and biological functions of terpenoids in plants. Advances in Biochemical Engineering/ Biotechnology. 2015;**148**:63-106

[52] Van Loon LC, Rep M, Pieterse CMJ. Significance of inducible defense-related proteins in infected plants. Annual Review of Phytopathology. 2006;**44**:135-162

[53] Hammerschmidt R. Phytoalexins: What have we learned after 60 years? Annual Review of Phytopathology. 1999;**37**:285-306

[54] Mithöfer A, Boland W. Plant defense against herbivores: Chemical aspects. Annual Review of Plant Biology. 2012;**63**:431-450

[55] Walling LL. The myriad plant responses to herbivores. Journal of Plant Growth Regulation. 2000;**19**(2):195-216

[56] Howe HF, Smallwood J. Ecology of seed dispersal. Annual Review of Ecology and Systematics. 1982;**13**:201-228. DOI: 10.1146/annurev.es.13.110182.001213

[57] Agati G, Azzarello E, Pollastri S, Tattini M. Flavonoids as antioxidants in plants: Location and functional significance. Plant Science. 2012;**196**:67-76. DOI: 10.1016/j.plantsci.2012.07.014

[58] Gill SS, Tuteja N. Reactive oxygen species and antioxidant machinery in abiotic stress tolerance in crop plants. Plant Physiology and Biochemistry. 2010;**48**(12):909-930. DOI: 10.1016/j.plaphy.2010.08.016

[59] Weston LA, Duke SO. Weed and crop allelopathy. Critical Reviews in Plant Sciences. 2003;**22**(3-4):367-389. DOI: 10.1080/713610861

[60] Howe GA, Jander G. Plant immunity to insect herbivores. Annual Review of Plant Biology. 2008;**59**:41-66. DOI: 10.1146/annurev.arplant.59.032607.092825

[61] Pieterse CMJ, Van der Does D, Zamioudis C, Leon-Reyes A, Van Wees SCM. Hormonal modulation of plant immunity. Annual Review of Cell and Developmental Biology. 2012;**28**:489-521. DOI: 10.1146/annurev-cellbio-101011-155754

[62] Peña-Cortés H et al. Ethylene and plant defense. Frontiers in Plant Science. 2016;7:991. DOI: 10.3389/fpls.2016.00991

[63] Thaler JS et al. The role of hormones in plant defense responses. Plant Biology. 2012;**14**:505-513. DOI: 10.1111/j.1438-8677.2012. 00608.x

[64] Barberis M, Calabrese D, Galloni M, Nepi M. Secondary metabolites in nectar-mediated plant-pollinator relationships. Plants. 2023;**12**(3):550. DOI: 10.3390/plants12030550

[65] Lepart J et al. Floral traits and pollinator preferences. Ecology. 2000;**81**:1715-1729

[66] Rogers HS, Donoso I, Traveset A, Fricke EC. Cascading impacts of seed disperser loss on plant communities and ecosystems. Annual Review of Ecology, Evolution, and Systematics. 2021;**52**:641-666

[67] Cotton TEA, Pétriacq P, Cameron DD, Meselmani MA, Schwarzenbacher R, Rolfe SA, et al. Metabolic regulation of the maize rhizobiome by benzoxazinoids. The ISME Journal. 2019;**13**:1647-1658

[68] Voges MJEEE, Bai Y, Schulze-Lefert P, Sattely ES. Plant-derived coumarins shape the composition of an Arabidopsis synthetic root microbiome. Proceedings of the National Academy of Sciences of the United States of America. 2019;**116**:12558-12565

[69] Graham PH. The efficiency of nodulation and nitrogen fixation in legume–Rhizobium symbiosis. Plant and Soil. 2008;**302**:1-7. DOI: 10.1007/s11104-007-9414-5

[70] Maffei ME, Gertsch J, Appendino G. Plant volatiles: production, function and pharmacology. Natural Product Reports. 2011;**28**(8):1359-1380

[71] Mönchgesang S, Strehmel N, Schmidt S, Westphal L, Taruttis F, Müller E, et al. Natural variation of root exudates in Arabidopsis thaliana—linking metabolomic and genomic data. Scientific Reports. 2016;**6**:29033

[72] Tsai HH, Schmidt W. Mobilization of iron by plant-borne coumarins. Trends in Plant Science. 2017;**22**:538-548

[73] Stringlis IA, de Jonge R, Pieterse CMJ. The age of coumarins in plant–microbe interactions. Plant & Cell Physiology. 2019;**60**:1405-1419

[74] Huang ACC, Jiang T, Liu YX, Bai YC, Reed J, Qu BY, et al. A specialized metabolic network selectively modulates Arabidopsis root microbiota. Science. 2019;**364**:eaau6389

[75] Kato-Noguchi H, Nakamura K, Okuda N. Involvement of an autotoxic compound in asparagus decline. Journal of Plant Physiology. 2018;**224**:49-55. DOI: 10.1016/j.jplph.2018.03.005

[76] Inderjit DSO. Ecological consequences of allelopathy. The Botanical Review. 2003;**69**:204-219. DOI: 10.1663/0006-8101(2003)069[0204: ECOA]2.0.CO2

[77] Sudatti DB, Duarte HM, Soares AR, Salgado LT, Pereira RC. New ecological role of seaweed secondary metabolites as autotoxic and allelopathic. Frontiers in Plant Science. 2020;**11**:347. DOI: 10.3389/fpls.2020.00347

[78] Kong CH, Li Z, Li FL, Xia XX, Wang P. Chemically mediated plant–plant interactions: allelopathy and allelobiosis. Plants. 2024;**13**(5):626. DOI: 10.3390/plants13050626